NUCLEAR REACTOR ENGINEERING
(PRINCIPLES & CONCEPTS)

For the Students of B.E./B.Tech. & M.E./M.Tech.

Dr. G. Vaidyanathan
B.Sc. Engg., MBA, Ph.D.
Visiting Professor, Department of Nuclear Science & Engineering,
SRM University,
Chennai
Former Director, Fast Reactor Technology,
Department of Atomic Energy,
Kalpakkam

S. CHAND & COMPANY PVT. LTD.
(AN ISO 9001 : 2008 COMPANY)
RAM NAGAR, NEW DELHI-110 055

S. CHAND & COMPANY PVT. LTD.
(An ISO 9001 : 2008 Company)
Head Office: 7361, RAM NAGAR, NEW DELHI - 110 055
Phone : 23672080-81-82, 9899107446, 9911310888 **Fax** : 91-11-23677446
Shop at: **schandgroup.com**; e-mail: **info@schandgroup.com**

Branches :

AHMEDABAD	:	1st Floor, Heritage, Near Gujarat Vidhyapeeth, Ashram Road, **Ahmedabad** - 380 014, Ph: 27541965, 27542369, ahmedabad@schandgroup.com
BENGALURU	:	No. 6, Ahuja Chambers, 1st Cross, Kumara Krupa Road, **Bengaluru** - 560 001, Ph: 22268048, 22354008, bangalore@schandgroup.com
BHOPAL	:	Bajaj Tower, Plot No. 2&3, Lala Lajpat Rai Colony, Raisen Road, **Bhopal** - 462 011, Ph: 4274723, 4209587. bhopal@schandgroup.com
CHANDIGARH	:	S.C.O. 2419-20, First Floor, Sector - 22-C (Near Aroma Hotel), **Chandigarh** -160 022, Ph: 2725443, 2725446, chandigarh@schandgroup.com
CHENNAI	:	No.1, Whites Road, Opposite Express Avenue, Royapettah, **Chennai** - 600014 Ph. 28410027, 28410058, chennai@schandgroup.com
COIMBATORE	:	1790, Trichy Road, LGB Colony, Ramanathapuram, **Coimbatore** -6410045, Ph: 2323620, 4217136 coimbatore@schandgroup.com **(Marketing Office)**
CUTTACK	:	1st Floor, Bhartia Tower, Badambadi, **Cuttack** - 753 009, Ph: 2332580; 2332581, cuttack@schandgroup.com
DEHRADUN	:	1st Floor, 20, New Road, Near Dwarka Store, **Dehradun** - 248 001, Ph: 2711101, 2710861, dehradun@schandgroup.com
GUWAHATI	:	Dilip Commercial (Ist floor), M.N. Road, Pan Bazar, **Guwahati** - 781 001, Ph: 2738811, 2735640 guwahati@schandgroup.com
HYDERABAD	:	Padma Plaza, H.No. 3-4-630, Opp. Ratna College, Narayanaguda, **Hyderabad** - 500 029, Ph: 27550194, 27550195, hyderabad@schandgroup.com
JAIPUR	:	1st Floor, Nand Plaza, Hawa Sadak, Ajmer Road, **Jaipur** - 302 006, Ph: 2219175, 2219176, jaipur@schandgroup.com
JALANDHAR	:	Mai Hiran Gate, **Jalandhar** - 144 008, Ph: 2401630, 5000630, jalandhar@schandgroup.com
KOCHI	:	Kachapilly Square, Mullassery Canal Road, Ernakulam, **Kochi** - 682 011, Ph: 2378740, 2378207-08, cochin@schandgroup.com
KOLKATA	:	285/J, Bipin Bihari Ganguli Street, **Kolkata** - 700 012, Ph: 22367459, 22373914, kolkata@schandgroup.com
LUCKNOW	:	Mahabeer Market, 25 Gwynne Road, Aminabad, **Lucknow** - 226 018, Ph: 4076971, 4026791, 4065646, 4027188, lucknow@schandgroup.com
MUMBAI	:	Blackie House, IInd Floor, 103/5, Walchand Hirachand Marg, Opp. G.P.O., **Mumbai** - 400 001, Ph: 22690881, 22610885, mumbai@schandgroup.com
NAGPUR	:	Karnal Bagh, Near Model Mill Chowk, **Nagpur** - 440 032, Ph: 2720523, 2777666 nagpur@schandgroup.com
PATNA	:	104, Citicentre Ashok, Mahima Palace , Govind Mitra Road, **Patna** - 800 004, Ph: 2300489, 302100, patna@schandgroup.com
PUNE	:	291, Flat No.-16, Ganesh Gayatri Complex, IInd Floor, Somwarpeth, Near Jain Mandir, **Pune** - 411 011, Ph: 64017298, pune@schandgroup.com **(Marketing Office)**
RAIPUR	:	Kailash Residency, Plot No. 4B, Bottle House Road, Shankar Nagar, **Raipur** - 492 007, Ph: 2443142,Mb.: 09981200834, raipur@schandgroup.com **(Marketing Office)**
RANCHI	:	Flat No. 104, Sri Draupadi Smriti Apartments, (Near of Jaipal Singh Stadium) Neel Ratan Street, Upper Bazar, **Ranchi** - 834 001, Ph: 2208761, ranchi@schandgroup.com **(Marketing Office)**
SILIGURI	:	122, Raja Ram Mohan Roy Road, East Vivekanandapally, P.O., Siliguri, **Siliguri**-734001, Dist., Jalpaiguri, (W.B.) Ph. 0353-2520750 **(Marketing Office)** siliguri@schandgroup.com
VISAKHAPATNAM	:	No. 49-54-15/53/8, Plot No. 7, 1st Floor, Opp. Radhakrishna Towers, Seethammadhara North Extn., **Visakhapatnam** - 530 013, Ph-2782609 (M) 09440100555, visakhapatnam@schandgroup.com **(Marketing Office)**

© 2013, Dr. G. Vaidyanathan

All rights reserved. No part of this publication may be reproduced or copied in any material form (including photo copying or storing it in any medium in form of graphics, electronic or mechanical means and whether or not transient or incidental to some other use of this publication) without written permission of the copyright owner. Any breach of this will entail legal action and prosecution without further notice.

Jurisdiction : All disputes with respect to this publication shall be subject to the jurisdiction of the Courts, tribunals and forums of New Delhi, India only.

First Edition 2013

ISBN : 81-219-4240-3   **Code** : 10 583

PRINTED IN INDIA
*By B.B. Press, A-37, Sector-67, Noida-201301 (U.P.)
and published by S. Chand & Company Pvt. Ltd., 7361, Ram Nagar, New Delhi -110 055.*

> "ENERGY IS THE ENGINE FOR
> ECONOMIC GROWTH
> AND WHEN INDIA MOVES AHEAD
> ON THE GROWTH PATH
> IT IS NECESSARY TO EXPLOIT
> EVERY ENERGY SOURCE
> AVAILABLE IN THE COUNTRY"
>
> *'Pt. Jawaharlal Nehru'*

Preface

India's Nuclear power programme was initiated in the 1950s by its founding father Dr. Homi Jehangir Bhabha. He had the view that if India were to become a developed nation, the per capita electricity consumption needed to go up. Among the major sources of electricity viz. fossil fuel, Hydel and nuclear, he felt that nuclear power would play a major role and laid the foundation for the setting up of nuclear power plants in India. The first plant with twin units was built at Tarapur in collaboration in USA. This was a boiling water reactor plant and needed enriched U235 upto 3%. Considering the fact that enrichment of uranium is a costly and complex technology to be developed in the country, it was decided to launch Pressurised Heavy Water Reactors (PHWR) based on the CANDU reactors developed in Canada. These required Heavy water for which technologies were available easily. The next 18 reactors were PHWRs. While all the reactors were of 220MWe, the two PHWRs built at Tarapur were 540 MWe. Today India has got into the construction of fast reactors, which would help in effective utilization of the natural uranium resources by converting fertile materials (U238 present as 99.3% in natural uranium and Th232 available in large quantities in India) into fissile materials Pu239 and U233.

The author has taught the subject of Nuclear Reactor engineering in the Training schools of the Department of Atomic Energy since 1980s. In 2010 the author superannuated after 37 years of service and became a visiting Professor in the Department of Nuclear Science and Engineering of SRM University, near Chennai. He taught Nuclear reactor design, Nuclear Thermal hydraulics and Nuclear safety to undergraduate students at the university. He was also a Visiting Professor at the Indian Institute of Technology, Chennai and Amity University, Noida for the M.Tech course in nuclear engineering. During the course of teaching, the author found that the students have to refer to a large number of books authored by foreign authors. Though many of these books, such as Nuclear Reactor Engineering by Samuel Glass-stone and Alexander Sesonske, and others were written few decades back, the language of these books were not suited for the Indian students. Also many books published were not giving enough exposure on the Pressurised Heavy Water Reactors (PHWR) and Fast Breeder Reactors (FBR). It may be noted that PHWRs form the first stage of India's nuclear programme, while FBRs form the second stage.

The above motivated the author to take up the work of writing a series of books. After careful examination of the syllabus of nuclear engineering in the different universities viz. Delhi, SRM and Amity, the IITs at Chennai and Kanpur, the first book was directed to Nuclear Reactor Engineering Concepts, as an introductory text for any nuclear engineering student.

Chapter 1 : Outlines the motivation for nuclear energy as compared to coal, oil and gas.

Chapter 2 : Deals with basics of reactor physics needed for the engineering student. Effort has been made to keep only the most relevant concepts and all terms used in connection with nuclear physics of reactors have been explained.

Chapter 3 : It is an introduction to the steam cycles and gas cycles used in nuclear and fossil fuelled plants. This was added in view of the fact that many students with a physics background join the M.Tech courses.

Chapter 4 : Takes the reader through all the facts of the nuclear fuel cycle from mining to fabrication, use in reactor to reprocessing and waste management.

Chapter 5 : Gives in brief the different components of a reactor, after giving a brief of the FERMI pile or the famous Chicago Pile.

Chapter 6 : Introduces the students to the basics of fluid flow and heat transfer concepts needed for the understanding of nuclear reactor operation.

Chapters 7 to 12 deal with brief description of Pressurised Water Reactor (PWR), Boiling Water Reactor (BWR), Pressurised Heavy Water Reactor (PHWR), Gas Cooled Reactors (GCR), Liquid Metal cooled Fast Breeder Reactor (LMFBR) and Molten Salt Reactor (MSR).

Chapter 13 : Dwells on India's nuclear power programme and the different capabilities built indigenously.

Chapter 14 : Takes the reader to the next generation reactors on which many countries are collaborating. This has indicated that the path of LMFBRs chosen by India to be the right one to meet the demands of large power.

Chapter 15 : Safety approaches at the design, construction and operation stages form the contents of this chapter. Examples of regulatory practices in India are highlighted.

Chapter 16 : On direct energy conversion of radiation into electricity using thermoelectric and thermionic generators, which are used for spacecrafts to supply power for long periods.

Chapter 17 : On Fusion, where brief concepts of Fusion and its advantages and issues in development have been brought out.

An appendix on Radiation and health has been added in the end to make the reader aware of radiation and its effects.

The preparation of this book, took the author through many books written in the 1940s to the 70s and these were a source of enriching his knowledge. Many books including those by Samuel Glasstone and Alexander Sesonske, El Wakil, Openshaw Taylor, Stephenson, Waltar and Reynolds were a source of information and inspiration in the preparation of this book. The author salutes these authors and many others whom could not be acknowledged due to constraints on space. This book contains information obtained from authentic and highly regarded sources. Reasonable efforts have been made to publish reliable data and information. In case of any errors readers may kindly bring to the notice of the author for corrections in future editions.

The author would like to acknowledge with thanks the support of many colleagues of the Department of Atomic Energy and Department of Nuclear Science and Engineering, SRM University, who gave useful comments and suggestions to improve the presentation.

Dr. G.Vaidyanathan
email: ganesan.vaidyanathan@gmail.com

Disclaimer : While the author of this book has made every effort to avoid any mistake or omission and has used his skill, expertise and knowledge to the best of their capacity to provide accurate and updated information. The author and S. Chand do not give any representation or warranty with respect to the accuracy or completeness of the contents of this publication and are selling this publication on the condition and understanding that they shall not be made liable in any manner whatsoever. S.Chand and the author expressly disclaim all and any liability/responsibility to any person, whether a purchaser or reader of this publication or not, in respect of anything and everything forming part of the contents of this publication. S. Chand shall not be responsible for any errors, omissions or damages arising out of the use of the information contained in this publication. Further, the appearance of the personal name, location, place and incidence, if any; in the illustrations used herein is purely coincidental and work of imagination. Thus the same should in no manner be termed as defamatory to any individual.

Detailed Contents

1. MOTIVATION FOR NUCLEAR ENERGY 1–12
1.0 Introduction 1
1.1 The Role of Electricity 1
1.2 Disparities among Countries 2
1.3 Sources of Energy 2
1.4 Present Sources of Electricity 3
1.5 Problems with Fossil Fuels 4
1.6 Resource Estimates (Table 1.3) 5
1.7 Global Climate Change 5
    1.7.1 Production of Carbon Dioxide 5
    1.7.2 The Effects of Greenhouse Gases 6
1.8 Alternatives to Fossil Fuels 6
    1.8.1 The Potential Role of Nuclear Energy 7
1.9 The Example of France 7
1.10 The Status of Nuclear Energy 8
1.11 Indias Energy Resources 9

2. BASIC PHYSICS OF NUCLEAR REACTORS 13–38
2.0 Introduction 13
2.1 Isotopes 13
2.2 Binding Energy 14
2.3 Nuclear Stability 15
2.4 Neutron Reactions 16
2.5 Radioactive Decay Law 17
2.6 Units of Radioactivity 18
2.7 Fission 18
    2.7.1 Fission Energy 18
    2.7.2 Critical Mass 18
    2.7.3 Gabon- The Natural Reactor 19
    2.7.4 Liquid Drop Model of Fission 19
2.8 Cross Section 20
2.9 Prompt And Delayed Neutrons 21
2.10 Neutron Life Cycle (Fig 2.6) 22

2.11	Infinite Multiplication Factor (K_∞)		22
	2.11.1 Four Factor Formula		23
		2.11.1.1 Fast Fission Factor (ε)	23
		2.11.1.2 Resonance Escape Probability (p)	24
		2.11.1.3 Thermal Utilization Factor (f)	24
		2.11.1.4 Reproduction Factor (η)	24
2.12	Effective Multiplication Factor (K_{eff})		25
	2.12.1 Fast Neutron Non-Leakage Probability (L_f)		25
	2.12.2 Thermal Neutron Non-Leakage Probability (L_t)		25
	2.12.3 Six Factor Formula		26
2.13	Neutron Moderation		26
2.14	Neutron Slowing Down		26
2.15	Neutron Moderators		27
2.16	Criticality and Reactor Power		28
2.17	Burners, Converters, and Breeders		28
2.18	Decay Heat		29
2.19	Reactivity		30
	2.19.1 Feedback Reactivity		31
		2.19.1.1 Moderator Temperature Coefficient	31
		2.19.1.2 Fuel Temperature Coefficient	31
		2.19.1.3 Pressure Coefficient	32
		2.19.1.4 Void Coefficient	33
		2.19.1.5 Neutron Poison	33
2.20	Reactor Kinetics		34
	2.20.1 Effective Delayed Neutron Fraction		35
2.21	Reactor Control		36
2.22	Biological Effects of Radiation and Shielding		36

3. BASICS OF POWER PLANT   39 – 48

3.0	Introduction	39
3.1	Carnot Cycle	40
3.2	Rankine Cycle	42
3.3	Reheat Cycle	43
3.4	Regenerative Cycle (Feed Water Heating)	45
3.5	Reheat-Regenerative Cycle	45
3.6	Brayton Cycle	46

4. REACTOR FUEL CYCLE   49 – 67

4.0	Introduction	49
4.1	Material Balance in the Nuclear Fuel Cycle	50
4.2	Uranium Mining	50
4.3	Uranium Milling	51

Contents

4.4	Conversion to Uranium Hexafluoride	53
4.5	Enrichment	53
	4.5.1 Gaseous Diffusion	53
	4.5.2 Gas Centrifuge	54
4.6	Fuel Fabrication	55
4.7	Fuel in Power Generation	55
4.8	Transport of Radioactive Materials	55
4.9	Spent Fuel Storage	56
4.10	Reprocessing	57
	4.10.1 Solvent Extraction	57
	4.10.1.1 Purex	57
	4.10.1.2 Urex	58
	4.10.1.3 Truex	58
	4.10.1.4 Diamex	58
	4.10.2 Pyroprocessing	59
	4.10.2.1 Electrolysis	59
4.11	Waste Management	60
	4.11.1 Types of Radioactive Wastes	60
	4.11.1.1 Exempt Waste and Very Low Level Waste	60
	4.11.1.2 Low-Level Waste	60
	4.11.1.3 Intermediate-Level Waste	61
	4.11.1.4 High-Level Waste	61
	4.11.2 Treatment and Conditioning of Nuclear Wastes	61
	4.11.2.1 Incineration	62
	4.11.2.2 Compaction	62
	4.11.2.3 Cementation	62
	4.11.2.4 Vitrification	63
4.12	Waste Disposal Methods	64
	4.12.1 Near-Surface Disposal	64
	4.12.2 Deep Geological Disposal	64
	4.12.3 Disposal in Outer Space	66
	4.12.4 Deep Boreholes	66
	4.12.5 Disposal at Sea	66

5. COMPONENTS OF A NUCLEAR REACTOR   68 – 76

5.0	Introduction	68
5.1	Fermi Pile	68
	5.1.1 Control	69
	5.1.2 Safety	69
	5.1.3 Radiation Monitoring	69
5.2	Reactor Core	69
5.3	Coolant	70

5.4		Control Rods	70
5.5		Moderator	71
5.6		Other Core Components	71
5.7		Containment (Fig 5.4)	71
5.8		Core Catcher	71
5.9		Steam Generator	71
5.10		Turbine Generator	72
5.11		Steam/Water System	72
5.12		Fuel Handling	72
5.13		Spent Fuel Cooling	72
5.14		Emergency Core Cooling	72
5.15		Types of Nuclear Reactors	72
	5.15.1	Pressurised Water Reactor(PWR)	72
	5.15.2	Boiling Water Reactor (BWR)	73
	5.15.3	Canada Deuterium-Uranium Reactors (CANDU)	74
	5.15.4	Sodium Cooled Fast Reactor(SFR)	74
	5.15.5	Advanced Gas Cooled Reactors (AGR)	75

6. REACTOR THERMAL HYDRAULICS 77 – 103

6.1		Reactor Heat Generation	77
	6.1.1	Fission Energy in Reactor	77
	6.1.2	Heat Generation after Shutdown	78
	6.1.3	Heat Generation in the Moderator	78
	6.1.4	Heat Generation in Reflectors and Shields	79
	6.1.5	Heat Generation in Structures	79
6.2		Heat Transport in the Fuel Element	79
	6.2.1	General Heat Conduction Equation	79
	6.2.2	Initial and Boundary Conditions	81
	6.2.2.1	Boundary Conditions (BC)	81
	6.2.2.2	Initial Condition	82
6.3		Heat Conduction in Cylindrical Fuel Rod	82
6.4		Thermal Properties	83
6.5		Temperature Distribution in Restructured Fuel	86
	6.5.1	Heat Conduction in Hollow Fuel Rod (Annular Fuel)	86
6.6		Gap Conductance	88
6.7		Heat Transport to the Coolant—Single Phase	88
	6.7.1	Fluid Flow Characteristics of Coolant	89
	6.7.2	Heat Transfer to Coolant	89
6.8		Heat Transfer Coefficients in Free Convection	92
6.9		Heat Transfer in Boiling	93
6.10		Pressure Drop in Reactor System	95
6.11		Two Phase Pressure Drop	97

		6.11.1 Flow Instability	98
		6.11.1.1 Static Instabilities (Fig 6.18)	99
		6.11.1.2 Dynamic Instability	99
6.12	Choked Flow		100
6.13	Condensation Heat Transfer		101
6.14	Pressure Drop in Rod Bundles		101

7. PRESSURIZED WATER REACTORS 104 – 123

7.0	Introduction	104
7.1	Reactor Configurations	105
	7.1.1 Overall System (Fig 7.1)	105
	7.1.2 Coolant	105
	7.1.3 Moderator	106
	7.1.4 Core Configuration	107
	7.1.5 Reactor Vessel	109
7.2	Reactor Control	110
7.3	Steam Generation	110
7.4	Other Primary System Components	113
	7.4.1 Primary Coolant Pumps	113
	7.4.2 Pressurizer	114
7.5	Auxiliary Systems	115
7.6	Steam Turbine Cycle	116
7.7	Chemical Volume Control System (Fig. 7.13)	117
7.8	Residual (Shutdown) Heat Removal Circuit	117
7.9	Emergency Core Cooling Circuit	119
7.10	Containment Systems (Fig 7.17)	120
	7.10.1 Containment Spray System	121
	7.10.2 Hydrogen Control in Containment	121
7.11	Advantages Over BWR	122
7.12	PWR Typical Issues	122

8. BOILING WATER REACTOR 124 – 133

8.0	Introduction	124
8.1	BWR Reactor Vessel Assembly	125
8.2	Fuel and Control Assemblies	126
8.3	Reactor Water Cleanup System (Fig 8.5)	127
8.4	Shutdown/Decay Heat Removal (Fig. 8.6)	127
8.5	Reactor Core Isolation Cooling (Fig. 8.7)	127
8.6	Standby Liquid Control System (Fig 8.8)	129
8.7	Emergency Core Cooling Systems (Fig. 8.9)	130
8.8	Boiling Water Reactor Containments	131

9. PRESSURISED HEAVY WATER REACTOR — 134 – 146

- 9.0 Introduction — 134
- 9.1 Genesis of CANDU HWR — 135
- 9.2 Reactor — 136
- 9.3 Moderator Systems — 136
- 9.4 Heat Transport Systems — 137
- 9.5 Feed and Bleed Circuit — 139
- 9.6 Fuel — 139
- 9.7 Fuel Handling — 140
- 9.8 Reactor Power Control — 141
- 9.9 Reactor Safety — 141
    - 9.9.1 Shutdown Systems (Fig 9.7) — 142
    - 9.9.2 Shutdown Cooling System — 142
        - 9.9.2.1 Emergency Core Cooling System (Fig 9.8) — 143
- 9.10 Containment System (Fig 9.9) — 143
- 9.11 Main Steam And Feedwater Systems — 144
    - 9.11.1 Steam Generator Pressure Control — 145
    - 9.11.2 Steam Generator Level Control — 145

10. GAS COOLED REACTOR — 147 – 152

- 10.0 Introduction — 147
- 10.1 MAGNOX-General description — 147
- 10.2 MAGNOX Reactors in Other Countries — 149
- 10.3 UNGG Reactors — 150
- 10.4 Advanced Gas Cooled Reactor (AGR) — 150

11. LIQUID METAL FAST REACTORS — 153 – 172

- 11.0 Introduction — 153
- 11.1 Flexible Use of Actinides — 155
- 11.2 Waste Minimization — 155
- 11.3 Overview of Sodium Cooled Fast Reactor (SFR) — 156
- 11.4 Layout of SFR Components — 158
    - 11.4.1 Pool Concept — 159
    - 11.4.2 Loop Concept — 160
- 11.5 Fuel Design — 160
    - 11.5.1 Fuel Element — 160
    - 11.5.2 Fuel Subassembly — 162
        - 11.5.2.1 Fuel Handling — 163
- 11.6 Intermediate Circuits — 163
    - 11.6.1 Intermediate Heat Exchanger (IHX) (Fig 11.9) — 164
    - 11.6.2 Steam Generators — 165
- 11.7 Sodium Pumps — 168

Contents

	11.7.1 Electromagnetic Pump	168
	11.7.2 Centrifugal Pumps (Fig 11.14)	169
11.8	Auxiliary Circuits	170
	11.8.1 Inert Gas System	170
	11.8.2 Trace Heating	170
	11.8.3 Sodium Purification System	171

12. MOLTEN-SALT REACTORS 173 – 177
- 12.0 Introduction — 173
- 12.1 Ornl Developmental Work — 173
- 12.2 Molten Salt Reactor Experiment (MSRE) — 174
- 12.3 Problems of MSR — 176
- 12.4 Impact of Lmfbr Programme on MSR — 176

13. INDIA'S NUCLEAR POWER PROGRAMME 178 – 190
- 13.0 Introduction — 178
- 13.1 Setting Up of Nuclear Establishment — 178
- 13.2 Research Reactor Apsara — 180
- 13.3 Canada India Research Reactor (CIRUS) — 180
- 13.4 Spent Fuel Reprocessing — 180
- 13.5 India's Nuclear Facilities — 181
  - 13.5.1 Power Reactors — 181
  - 13.5.2 Research Reactors — 182
  - 13.5.3 Uranium Enrichment Facilities — 182
  - 13.5.4 Heavy Water Production — 182
  - 13.5.5 Fuel Fabrication — 183
  - 13.5.6 Reprocessing Facilities — 183
  - 13.5.7 Uranium Mining, Milling — 183
- 13.6 Future Indian Stage 3 Programme — 183
  - 13.6.1 Advanced Heavy Water Reactor (AHWR) — 184
  - 13.6.2 Accelerator Driven Subcritical Systems (ADS) — 186
  - 13.6.3 Compact High Temperature Reactor (CHTR) — 187

14. NEXT GENERATION OF REACTORS 191 – 203
- 14.0 Introduction — 191
- 14.1 Generation I — 191
- 14.2 Generation II — 192
- 14.3 Generation III — 192
  - 14.3.1 ABWR — 192
- 14.4 Generation III+ — 194
  - 14.4.1 EPR — 195
- 14.5 Generation IV Concepts — 195

14.5.1	Gas-Cooled Fast Reactor (GFR)	196
14.5.2	Lead-Cooled Fast Reactor (LFR)	197
14.5.3	Molten Salt Reactor (MSR)	198
14.5.4	Sodium-Cooled Fast Reactor (SFR)	198
14.5.5	Supercritical Water-Cooled Reactor (SCWR)	200
14.5.6	Very High Temperature Reactor (VHTR)	201

14.6 Actinide Management — 201
14.7 Kick-Starting the Hydrogen Economy — 202

15. SAFETY APPROACHES IN REACTOR DESIGN   204 – 217

15.0 Introduction — 204
15.1 Comprehensive Safety Analysis — 204
    15.1.1 Defence in Depth — 205
    15.1.2 The Fundamental Safety Functions — 206
15.3 Current Safety Approach — 206
    15.3.1 Achieving Design Safety in Practice — 207
        15.3.1.1 Facilities to Control & Shutdown Reactor (Fig 15.2) — 207
        15.3.1.2 Facilities to cool the reactor — 208
    15.3.2. Containment (Fig 15.3) — 208
    15.3.3 Facilities to Support Safety Facilities — 208
        15.3.3.1 Redundancy Diversity and Independence — 208
15.4 Design Basis Events (DBE) — 209
    15.4.1 Event Classes — 210
        15.4.1.1 Class I-Normal Operation and Operational Transients — 210
        15.4.1.2 Class II Events of Moderate Frequency — 211
        15.4.1.3 Class III Events of Low Frequency — 211
        15.4.1.4 Class IV Multiple Failures and Rare Events — 211
    15.4.2 Beyond Design Basis Events — 211
15.5 Probabilistic Safety Analysis — 212
15.6 Regulatory Process In India — 213
    15.6.1 Site Approval — 213
    15.6.2 Construction Approval — 213
    15.6.3 Operating License — 214
    15.6.4 Regulatory Inspection — 214
15.7 Radiation Dose Limits — 214
15.8 Radiation Exposure — 214

16. DIRECT ENERGY CONVERSION   218 – 227

16.0 Introduction — 218
16.1 Thermoelectric Generators — 218
    16.1.1 Radionuclide Thermoelectric Generators — 219
    16.1.2 Reactor Thermoelectric Generators — 221

16.2	Thermionic Electrical Generators	222
	16.2.1 Radionuclide Thermionic Generators	223
16.3	AMTEC Conversion	224
16.4	Betavoltaic Batteries	224
16.5	Radioisotopes for Thermal Power Sources	225

17. FUSION 228 – 236

17.0	Introduction	228
17.1	Binding Energy	228
17.2	Fusion Reactions	230
17.3	Fusion Fuel Availability	230
17.4	Issues In Fusion	231
17.5	Thermonuclear Reaction In Plasma	232
	17.5.1 Inertial Confinement Fusion (Icf)	232
	17.5.2 Magnetic Confinement	233
17.6	Lawson's Criteria	233
17.7	Fusion Power Plants	234
17.8	Safety And Environmental Considerations	234
17.9	The "Cold Fusion" Confusion	234
17.10	Fusion Research In India	235
17.11	Advantages Of Fusion Energy	235

APPENDIX: RADIATION AND HEALTH   237 – 248

1

Motivation for Nuclear Energy

1.0 INTRODUCTION

The exploitation of new sources of energy has been central to human progress from the early Stone age to today's technological world to fulfill the need for survival. The first step was learning to make fire and then to control fire, with wood or other biomass as the fuel. This was followed by harnessing wind for ships and windmills, the use of river water for power generation, and much later, the exploitation of chemical energy from the burning of coal, oil, and natural gas. Nuclear energy, which first emerged in the middle of the 20th century, is the latest energy source to be used on a large scale. Technological advances however do inflict damage on the environment, whatever is the sphere of activity. Nevertheless, most people in the developed countries gladly accept the fruits of technological advances, which the people in the developing countries aspire. While the burden of inefficient or unnecessary energy consumption may be reduced, it is unlikely that there will be a consensus favoring a substantial reduction in energy use in most of the developed countries or a voluntary stemming of the rise of energy use in the developing ones, with their growing population looking for improved living standards. Thus, the world will demand increasing supplies of energy in the coming years. Nuclear power provides one option for supplying this energy.

1.1 THE ROLE OF ELECTRICITY

Overall, in the 20th century, electrification has been almost synonymous with modernization. It has changed the mechanics of the home, with convenient lighting, refrigeration, and motor-driven appliances, plus expanded entertainment and cultural resources. In industry, electrical motors allow machines to be used where and when they are needed and electric equipment can deliver heat in highly controllable forms—for example, with electric arcs, laser beams, and microwaves. There is increasing use of electricity in agriculture as well. Medical diagnosis and treatment has been transformed by use of equipment ranging from X-rays and high-speed dental drills to lasers and magnetic resonance imaging. Electricity has made possible entirely new modes of communication, as well as the development of computers and the associated means for exchanging and storing information.

Electricity plays a central role in virtually all spheres of technological life excluding transportation as of now. Considering the fact that there is a growth in rapid electric trains in Japan and Europe and introduction of electric trains in metros in all developing countries, the demand for electricity is on the rise. In the industrialized countries, as represented by the Organization of Economic Cooperation and Development (OECD), electricity consumption rose by 117% in the 1973–2001 period. This far outstripped the growth, during the same period, in population (26%) and in total primary energy supply (42%). The developing countries true to their description, lagged for many years in the use of electric power, but there has been rapid recent growth in some countries. In China, electricity consumption rose from 69 gigawatt years (GW yr) in 1991 to 138 GW yr in 2000, an average annual increase of 8.1%. The growth rate in South Korea in this period was still higher, averaging 10.6% per year. It appears inevitable that the demand for electricity will increase on a worldwide basis, even if conservation and energy efficiency restrains the growth in some countries. This increase will be driven by

(a) increasing world population,
(b) increased per capita use of energy, at least in successful developing countries, and
(c) an increase in electricity's share of the energy budget due to the convenience of electricity in some applications, its uniqueness in others, and its cleanliness.

Additional demands for electricity may arise for the production of hydrogen and for the desalination of seawater.

1.2 DISPARITIES AMONG COUNTRIES

The disparities among countries are great. In 2001, the per capita consumption of energy for industrialized countries such as France and Japan was about 14 times that of India and almost 50 times that of Bangladesh. On the other hand the figure for USA is twice that of Japan and France. It might be possible for the United States to reduce its per capita energy use, but in many countries there is a need for more energy. The, U.S. energy consumption per capita hardly changed from 1980 to 2001, whereas per capita consumption more than doubled for India and rose over 150% for Bangladesh—a considerable accomplishment, especially considering the substantial population growth in these countries during the same period. To accommodate an increasing population and an increased per capita demand in much of the world, world energy production may have to more than double over the next 50 years. If present trends continue, most of this additional energy will come from fossil fuels.

1.3 SOURCES OF ENERGY

For well over 100 years, the dominant energy sources in the industrialized world have been fossil fuels—coal, oil, and natural gas—and these now dominate in most of the developing world as well. Other major contributors, of varying importance in different countries, include hydroelectric power, nuclear power, and biomass.

Table 1.1 indicates the main sources of energy for the United States and the world. The total energy generated in 2001 was $1.16*10^8$ Gwh and 25 % was the share of USA alone. The dominance of fossil fuels is brought out in these data. They provide 86% of the primary energy for both the United States and the world. The remainder is divided between nuclear and renewable sources. The magnitude of biomass consumption is difficult to establish accurately, because much of it involves the collection of wood and wastes on an individual or small-scale basis, outside of commercial channels. Its use is therefore less well documented than is the use of fuels that are purchased commercially. The problem with biofuels is the low energy density, a term that refers to the energy that can be harnessed from a given area. Energy analysts estimate that the power density of corn ethanol to be as low as 0.05 w/sq. m compared to 28 w/sq.m for natural gas. The important renewable source is hydroelectric power.

Table 1.1: Main Sources of world energy

Source	World 2001 %	USA 2001 %	India 2010 %
Fossil Fuels			
Petroleum	30	30	8.4
Coal	24	23	53.4
Natural Gas	23	24	11
All Fossil Fuels	86	86	72.8
Renewable Sources			
Hydroelectric	6.6	2.6	18
Other Renewable	0.8	3.3	6.91
All Renewable	7.4	6.0	24.91
Nuclear	6.5	8.3	2.3
Total	100	100	100
Total Energy (Gwh)	$1.16*10^8$	$3.87*10^7$	$0.91*10^6$

Table 1.2: Power Generation India (2010)

Source	Generation Capacity (Mw)
Non Renewables	
Thermal	117975
Hydro	36863
Non Utilities	28474
Nuclear	4560
Total	187872
Renewables	
Wind Power	11806
Biomass	2199
Waste energy	65.0
Mini Hydel	2735
Solar Power	10.28
Total	16815.28

1.4　PRESENT SOURCES OF ELECTRICITY

Most electricity is now being generated by the combustion of fossil fuels. World generation in 2001 was roughly two-thirds from fossil fuels and one-sixth each from hydroelectric and nuclear power. The electricity generated by renewable sources other than hydroelectric power is not always reliably reported, because it is in large measure produced by entities other than utilities and the accounting is not as reliable as for utility generation. These sources include biomass energy (wood and wastes), geothermal energy, wind energy, and direct forms of solar generation. The present (2010) sources of electricity in India are given in table 1.2.

1.5 PROBLEMS WITH FOSSIL FUELS

The long-standing impetus for the development of nuclear power has been the eventual need to replace fossil fuels—oil (or petroleum), coal, and natural gas. Their supply is finite and eventually the readily available resources will be consumed, at different rates for the different fuels. Concern over oil was less visible in the 1990s than it had been in the late 1970s, because of new findings. However, if oil shortages have been deferred, they cannot in the long run be avoided. Known and projected resources of oil are heavily concentrated in the Persian Gulf region, and unless substitutes for oil are found, the world will face a continuing series of economic and political crises as countries compete for the dwindling supplies.

Natural gas resources may exceed those of oil, measured in terms of total energy content, and the present world consumption rate is less for gas than for oil. Therefore, global shortages are somewhat less imminent. Nonetheless, gas is also a limited resource, with reliance on unconventional resources speculative. Besides it is a valuable feedstock for chemical industries. Coal resources are much more plentiful than those of either oil or natural gas, but coal is the least environmentally desirable. Coal became important in the 17th century, and it is now the leading fuel. It has not had a clean history, with chronic pollution punctuated by severe incidents such as 4000 deaths in the London smog of 1952. However, output of chemical pollutants from coal, particularly sulphur dioxide, can be greatly reduced by "cleaner" burning of the coal, at a moderate additional cost. Coal mining continues to be a hazardous operation.

The production of carbon dioxide in the combustion of fossil fuels presents a more difficult problem. Unless much of this carbon dioxide can be captured, the resulting increase in the concentration of carbon dioxide in the atmosphere carries with it the possibility of significant global climate change.

A fact that has gone unnoticed is the release of radioactivity from coal fired power plants. Coal mined from under the ground does contain traces of uranium and thorium. For a large number of coal samples, according to U.S. Environmental Protection Agency figures released in 1984, average values of uranium and thorium content have been determined to be 1.3 ppm and 3.2 ppm, respectively. As the coal burns, the uranium and thorium particles get out along with the flue gases into the environment. The main sources of radiation released from coal combustion include not only uranium and thorium but also daughter products produced by the decay of these isotopes, such as radium, radon, polonium, bismuth, and lead. Naturally occurring radioactive potassium-40 is also a significant contributor.

For comparison, according to National Council on Radiation Protection and Measurements (NCRP) reports, population exposure from operation of 1000-MWe nuclear and coal-fired power plants amounts to 490 man-rem/year for coal plants and 4.8 man-rem/year for nuclear plants. Thus, the population effective dose equivalent from coal plants is 100 times that from nuclear plants. For the complete nuclear fuel cycle, from mining to reactor operation to waste disposal, the radiation dose is cited as 136 man-rem/year; the equivalent dose for coal use, from mining to power plant operation to waste disposal, is not documented.

Among the fossil fuels, natural gas has significant advantages. It is the least environmentally damaging, in terms of both chemical pollutants and carbon dioxide production. Nonetheless, reliance on natural gas as more than a short-term stop gap involves two significant uncertainties or problems. First, gas supplies may be limited to the standard conventional resources, advancing the time at which the availability of gas will become a problem and prices will rise substantially. Second, although preferable to coal in this regard, natural gas is still a source of greenhouse gases, primarily carbon dioxide from combustion and secondarily methane from leaks.

1.6 RESOURCE ESTIMATES (TABLE 1.3)

Almost 30% of the world's oil has come from the Persian Gulf region in recent years. As other countries gradually use up their resources, the abundant reservoirs of the Persian Gulf will become proportionally more important, further increasing the political sensitivity of the region. USA in 2002 relied on (net) imports for 53% of its oil supply. The United States and world dependence on oil from the Persian Gulf region has led to considerable political and military unrest. In response, there have been intermittent attempts by the U.S. government to lessen the country's dependence on oil imports. The potential for increased domestic supplies is limited, and the most promising avenue is reduced consumption.

Table 1.3: Estimates of World Resources of Fossil Fuels

	Resource Base(EJ) Conventional	Resource Base (EJ) Unconventional	Consumption (EJ/year)
OIL	$12.1*10^3$	$20.3*10^3$	142
NATURAL GAS	$16.6*10^3$	$33.2*10^3$	84
COAL	$200*10^3$		92

* The exajoule (EJ) is equal to 10^{18} joules.

However, two-thirds of all petroleum use in the United States is for transportation, and change cannot be accomplished quickly because the living patterns depend heavily on automobiles and trucks. An effective approach for the near term would be to increase the average fuel efficiency (in miles per gallon) of conventional vehicles. Further gains could be made by increased use of mass transportation, especially electrified mass transportation. The replacement of petroleum-based fuels by alternatives, particularly hydrogen, is a possibility for the further future. Here, nuclear power could play a role as an energy source for hydrogen production.

1.7 GLOBAL CLIMATE CHANGE

1.7.1 Production of Carbon Dioxide

If the Earth had no atmosphere, its average surface temperature would be about −18°C. The Earth is kept at its relatively warm temperature by molecules in the atmosphere, including water molecules and carbon dioxide molecules that absorb some of the infrared radiation emitted by the Earth and prevent its escape from the Earth's environment. This is the natural "greenhouse effect." Since the beginning of the industrial era, additional gases have been emitted into the atmosphere—particularly carbon dioxide (CO_2)—which adds to this absorption and are believed to further increase the Earth's temperature. This increment is referred to as the anthropogenic greenhouse effect. Warnings about the effects of CO_2 emissions date to the 19th century, but they have become a matter of widespread concern only since the 1970s. The anticipated consequences are described as "global warming" or, more broadly, as "global climate change."

The production of CO_2 is the inevitable accompaniment of any combustion of fossil fuels. The amount released per unit energy output varies for the different fuels, due largely to differences in their hydrogen content. Natural gas is primarily methane (CH_4) and a considerable fraction of its combustion energy comes from the chemical combination of hydrogen and oxygen. Its ratio of carbon dioxide production to energy production is the lowest among the fossil fuels.

The releases are usually specified in terms of the mass of carbon (C), not CO_2. Other greenhouse gases include methane, chlorofluorocarbons, and nitrous oxide. For our purpose, approximate average values are adequate. Approximate coefficients, in mega tonnes (Mt) of carbon per Exa joule (EJ) of energy, are 24.6 Mt/EJ for coal, 18.5 Mt/EJ for petroleum, and 13.7 Mt/EJ for natural gas. These numbers illustrate the benefit of switching from coal to natural gas, when possible. (1 EJ is equal to 10^{18} J)

1.7.2 The Effects of Greenhouse Gases

The increases during the past century in the atmospheric concentrations of greenhouse gases, especially carbon dioxide, have been unambiguously established. The projected effects for the period until 2100 include the following:

- An increase in global average temperature on the Earth's surface of 1.4°C to 5.8°C (2.5–10.4°F). About one-half of this rise is anticipated to take place by 2050.
- Increased average global precipitation.
- A rise in the average sea level due to the melting of glaciers and the thermal expansion of the oceans, in the broad interval of 9–88 cm.
- Increased frequency and intensity of "extreme events," including "more hot days, heat waves, heavy precipitation events, and fewer cold days," with possibly "increased risks of floods and droughts in many regions".
- A potentially devastating, but also quite uncertain, possibility is the collapse of the West Antarctic Ice Sheet. If it occurs, it would cause a 5-m rise in sea level, affecting millions of people living in low-lying coastal regions.

The elimination or reduction of carbon dioxide emissions from coal-fired electricity generation is straightforward, if one is willing to pay the costs. It could be done by substitution with nuclear power or renewable sources for coal or, if the technique proves practical on a large scale, by sequestration of the carbon dioxide emitted from coal-fired power plants. The reduction of carbon dioxide emissions in transportation is more difficult. There is now considerable speculation about hydrogen as an energy carrier for use in transportation, but its practicality has not been established.

1.8 ALTERNATIVES TO FOSSIL FUELS

The challenge in energy policy is to reduce CO_2 emissions and the world's dependence on oil while satisfying a substantially increased demand for energy. Putting aside the still-speculative possibility of sequestering carbon dioxide, this challenge reduces to that of using energy more efficiently and finding substitutes for fossil fuels. These can be put into two broad categories:

Renewable sources. Most of these sources—including hydroelectric power, wind power, direct solar heating, photovoltaic power, and biomass—derive their energy ultimately from the Sun and will not be exhausted during the next billion years. Geothermal energy and tidal energy are also renewable, in this sense, although they do not rely on the sun. The expansion of hydroelectric power (which is substantially used) is constricted by limited sites and environmental objections, whereas wind (for which the resource is large) is as yet less used and thus is not fully proven as a large-scale contributor.

Nuclear sources. The two nuclear possibilities are fission and fusion. The latter would be inexhaustible for all practical purposes, but developing an effective fusion system remains an uncertain hope. Fission energy would also have an extremely long time span if breeder reactors are employed, but with present-day reactors limits on uranium (or thorium) resources could be an eventual problem. At present, fission power faces problems of public acceptance and economic competitiveness. The broad alternatives of renewable energy and nuclear energy can be considered as being in competition, with one or the other to be the dominant choice, or complementary, with both being extensively employed.

When the possibility of nuclear energy was first recognized in the 1930s and early 1940s, it had the attraction of offering very large amounts of energy from very small amounts of material. This excited the imagination of scientists and fission was looked upon as a very promising potential energy source. Following the technological success of the World War II atomic bomb program, it appeared likely that commercial nuclear energy would prove to be practical.

There is considerable disagreement today on both the immediate and ultimate potential of solar energy. One view is that it is not possible to obtain adequate amounts of energy from renewable sources, either now or in the predictable future. An opposing view holds that a combination of renewable and conservation could, in a matter of decades, make fossil fuels and nuclear energy unnecessary. Lacking confidence that renewable energy alone will suffice to replace fossil fuels, the U.S. government has adopted a policy in recent years of keeping the nuclear option alive, without a major investment in fostering its growth. However subsequent to 2010, there is a nuclear renaissance, in view of the energy demand and budget allocations to research in the nuclear arena are on the increase.

1.8.1 The Potential Role of Nuclear Energy

Here, we consider the contribution nuclear power could make to solve the world's energy problems. For the developed countries, where the increase in energy demand over the next 50 years could be fairly small if conservation measures are vigorously implemented, the most important contribution would be in direct displacement of fossil fuel sources. Potential measures include the following:

- the gradual replacement of present coal-fired power plants by nuclear plants. Both coal and nuclear plants are used primarily for base load generation; their roles are interchangeable.
- the use of nuclear power rather than natural gas when new capacity is needed. This would free natural gas to replace oil or coal in heating and other applications.
- the replacement of petroleum in transportation. As already discussed in the context of resources, this change is more difficult to implement. Looking ahead several decades, nuclear energy could contribute by providing power for electric vehicles, hydrogen production, and electrified mass transportation.
- the replacement of fossil fuels by electricity for heating. In industry, electric heating can be applied at the desired location and time, with unique precision. In homes and commercial buildings, efficient use of electricity can be achieved with heat pumps or controlled zone heating, although not with electric central furnaces.

For the developing countries, which hope to increase their energy consumption substantially, the expanded use of nuclear power faces the problem of limited capital resources and, in some cases, an inadequate technical base. However, the two largest countries in this category—China and India—have considerable nuclear sophistication and to the extent capital is available, they could turn to nuclear power instead of coal or natural gas to fuel their electricity expansion.

1.9 THE EXAMPLE OF FRANCE

The changes suggested above could not be implemented immediately, but a significant part could be accomplished on the time scale of decades, as illustrated by the history of the French energy economy since the early 1970s. Nuclear advocates cite this history as an example of the contribution nuclear energy can make in reducing carbon dioxide emissions and, in some situations, reducing the demand for oil. Between 1970 to 1995 —a period during which nuclear share of electricity generation rose from 6% to 77%, while the Gross Domestic Product (GDP) rose by 7.1%. However the population increased by 13% in this period. During this period electricity generation in France more than tripled and total energy supply rose 56%, while carbon emissions and petroleum use each dropped 16%. Some of these accomplishments can be attributed to the increased use of natural gas, but nuclear power deserves the lion's share of the credit. The replacement of fossil fuels by nuclear energy in the generation of electricity was particularly noteworthy. The fossil fuel share of electricity generation dropped from 62% to 8% in this period while the nuclear share increased from 6% to 77%. The drop in petroleum use also meant lower oil imports.

1.10 THE STATUS OF NUCLEAR ENERGY

General perceptions of nuclear energy, among both the public and policy makers, have undergone dramatic shifts in the past 50 years. As nuclear energy emerged in 1945 from scientific obscurity and military secrecy, it began to be talked of in speculative terms as an eventual power source. Within a decade, an enthusiastic vision developed of a future in which nuclear power would provide a virtually unlimited solution for the world's energy needs. It was not difficult to picture nuclear power as the ideal energy source. With the use of breeder reactor (where elements like uranium 238 and thorium 232 could be converted to fissile plutonium 239 and uranium 233), nuclear power would be ample in supply. As experience was gained in reactor construction, it would become economical; and because a nuclear reactor would emit virtually no pollutants, it would be clean.

There was also a negative side, as some doubters pointed out from the beginning. Very large amounts of radioactivity would be produced. In principle, practically none need escape, but the possibility of mishaps could not be totally excluded. Further, benign nuclear power had a malign older sibling in nuclear weapons. Although many people understood that a reactor itself could not explode like a bomb, some members of the public feared that in some way controlled nuclear energy might go out of control.

The optimists prevailed for two decades, into the early 1970s, and many nuclear reactors were designed, built, and put into operation in the United States and Europe. Part of the motivation for this development was the desire of countries to reduce their heavy dependence on oil, which they expected would eventually be in short supply. The first oil crisis came in 1973, even sooner than had been anticipated. Just as the nuclear buildup was gaining momentum, an oil embargo was imposed by the Organization of Petroleum Exporting Countries (OPEC), as a sequel to the October 1973 war between Egypt and Israel.

An immediate impulse was to rely even more on nuclear power as a substitute for oil. This was especially true in the United States, where nuclear energy had already appeared to many, as an important key to "energy self-sufficiency." However, the embargo had unanticipated effects. It focused new attention on the possibility of reducing all energy consumption, and the rising oil prices slowed the pace of economic growth. These factors sharply reduced the demand for electricity and, therefore, the pressure to add new nuclear power plants. At the same time, the economic costs of nuclear power and fears about nuclear power both began to grow. The Three Mile Island accident in 1979 and the Chernobyl accident in 1986 hit a world becoming more attuned to believing the worst about nuclear power. The recent accident at Fukushima in Japan after a Tsunami has once again brought in review of current designs. Such reviews are expected

Motivation for Nuclear Energy

to result in safer designs, making nuclear reactors safer.

Nuclear energy development was stopped or brought to a crawl in all but a few countries during the 1980s and 1990s. Contributing factors to this decline included a gradual reduction in oil and gas prices, rising nuclear costs, the sluggishness of the growth in energy demand and general fears of nuclear power. Nuclear power may seem dormant or dying in Western Europe, but it remains a vital activity in parts of Asia. The actual picture varies from country to country, with considerable long-term uncertainty in most industrialized countries.

There has also been widespread concern over the possibility of reactor accidents, the disposal of radioactive wastes, and the possibility that nuclear power could help some countries produce nuclear bombs. Since the terrorist attacks of September 11, 2001, there has also been heightened fear that nuclear facilities might be vulnerable targets for future attacks.

In counterpoint, there is also the widely expressed belief that these concerns are exaggerated and that it is less dangerous to use nuclear power than to try to get by without it. The goals of reducing CO_2 production and world dependence on oil will be hard to achieve under any circumstances. They will be all the harder to achieve without taking advantage of all effective energy sources—including, in this view, nuclear power. It is impossible to resolve these controversies in a universally convincing fashion. Instead of attempting such a task, the remainder of the book discusses basic aspects of nuclear power—primarily technical aspects. The emphasis will be on describing nuclear reactors and the nuclear fuel cycle and examining the associated issues.

1.11 INDIAS ENERGY RESOURCES

Our dream is to realise a quality of life for people commensurate with other developed countries. This needs generation of 5000 kWh per year per capita, and demands a total capacity of 7500 billion kWh per year for a population of 1.5 billion by 2050. These calls for a strategic growth in electricity generation considering: Energy resources, self sufficiency, Effect on local, regional & global environment, Health externalities, demand profile & energy import scenario.

Table 1.4: SOURCES OF ELECTRICITY GENERATION IN INDIA

	Amount	Thermal energy			Electricity potential
		EJ	TWh	GWYr	GWe-Yr
Fossil					
Coal	38 -BT	667	185,279	21,151	7,614
Hydrocarbon	12 -BT	511	141,946	16,204	5,833
Non-Fossil Nuclear					
Uranium-Metal	61,000 -T				
In PHWRs		28.9	7,992	913	328
In Fast breeders		3,699	1,027,616	117,308	42,231
Thorium-Metal	2,25,000 -T				
In Breeders		13,622	3,783,886	431,950	155,502
Renewable					
Hydro	150 -GWe	6.0	1,679	192	69
Non-conventional renewable	100 -GWe	2.9	803	92	33

Sources of electricity generation in India and their expected contributions are given in table

1.4. A recent study conducted by the Department of Atomic Energy, India, has indicated that it is necessary to have ¼ of total electricity generation through nuclear if we are to meet the energy demand in 2050. Nuclear energy will also need to play a progressively increasing role for non-grid-based-electricity applications (hydrogen generation, desalination, compact power packs). For a large country like India, with huge future energy requirements, depending largely upon import of energy resources and technologies is neither economically sustainable nor strategically sound for energy security.

India is presently following a three stage nuclear power programme in which the first stage is based on Pressurised Heavy Water Reactors (Figure 1.1). Based on the plutonium obtained from the reprocessing of the spent fuel from these reactors, the fast breeder reactor programme will form the second stage. The fast breeder reactors would produce Pu^{239} at a faster rate by converting the unused U^{238} in the spent fuel from the first stage of reactors. In the later stages they would be used for breeding U^{233} from Th^{232}. Thus a large base of nuclear plants involving fast breeders would be set up in the second stage. In the third stage thorium breeders based on the U^{233}-Pu^{239}-Th^{232} cycle would be set up. It is clear that to facilitate wide-spread and long term use of nuclear power a sustainable nuclear fuel strategy, based on closed nuclear fuel cycle (reprocessing to recover Pu^{239} and unburnt $Uranium^{235}$ which are fissile material) and thorium utilisation is essential. Taking cognisance of its resource position, the Indian priority for adopting this strategy has been high.

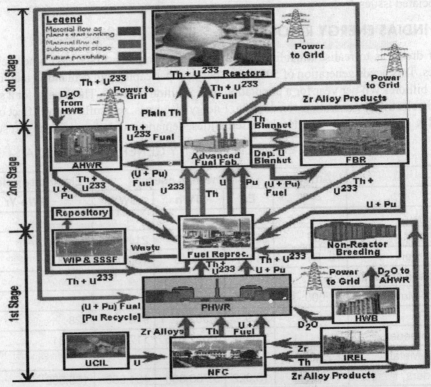

Fig. 1.1: India's Nuclear Power Programme

Study of the nuclear resources has indicated a gap of around 40 GWe after considering the indigenously available natural uranium resources. While the import of Nuclear power plants in sixties (Tarapur 1 & 2, RAPS 1 & 2) was with the objective of long term economics and

Motivation for Nuclear Energy

sustainability for building a large programme, the present philosophy is to import light water reactors to fill the gap in energy supply to meet the demands of electricity. A beginning has been made with the import of two Light Water reactors from Russia. The study has also indicated that introduction of imported reactors should be immediate with introduction of 40 GWe in the period 2012 to 2020 A delay by 10 years would increase the deficit of electricity to 178 GWe as can be seen from the figure 1.3.

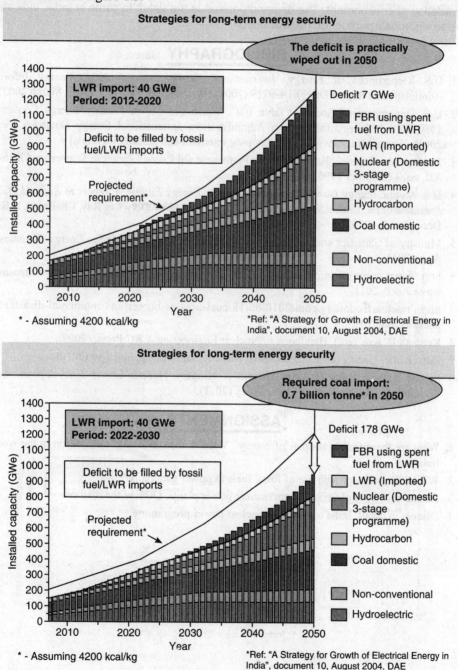

Fig. 1.3: Strategies for Long Term Energy Security

SUMMARY

This chapter has brought out the different energy sources for electricity generation and the role of nuclear power. With the demand for energy increasing with living standards of people in different parts of world, there is the need to use the different resources judiciously. Nuclear power could contribute to a large extent to fulfill the demand for power without the degradation of environment. The demand position in the years to come in India, indicate the need to vigorously pursue the nuclear option.

BIBLIOGRAPHY

1. U.S. Department of Energy, *International Energy Annual 2001*, Energy Information Administration Report DOE/EIA-0219 (2001) (Washington, DC: U.S. DOE, March 2003).
2. U.S. Department of Energy, "Table E1C. World Per Capita Primary Energy Consumption, 1980–2001," Energy Information Administration, International Energy Database (February 2003). [From: http://www.eia.doe.gov/pub/ international/ iealf/tablee1c.xls]
3. James J. MacKenzie, "Heading off the Permanent Oil Crisis," *Issues in Science and Technology* XII, no. 4 (Summer 1996): 48–54.
4. U.S. Atomic Energy Commission, *The Nation's Energy Future*, A Report to Richard M. Nixon, President of the United States, WASH-1281, submitted by Dixy Lee Ray, Chairman, U.S. AEC, December 1, 1973.
5. Ministry of Statistics and Programme Implementation, Govt. of India, Energy Statistics 2011, April, 2011.
6. http://www.scientificamerican.com/article.cfm?id=coal-ash-is-more-radioactive-than-nuclear-waste.April, 2011.
7. http://nuclearfissionary.com/2010/06/18/coal-creates-bigger-environmental-disasters-than-nuclear/,June 2010.
8. Kenneth D. Kok (ed.), Handbook of Nuclear Engineering, CRC Press, (2009).
9. Dan Gabriel Cacuci (ed.), Handbook of Nuclear Engineering, Springer (2010).
10. Steven B., Krivit, Jay H. Lehr, Thomas B. Kingery, Nuclear Energy Encyclopedia: Science, Technology, and Applications John Wiley (2011).

ASSIGNMENTS

1. What are the different sources of energy? What is their available distribution in the world and India?
2. What is the problem with use of fossil fuels in power generation?
3. France has >80% electricity from nuclear power plants. Discuss the reasons.
4. Discuss in detail, India's ambitious nuclear power programme.

2

Basic Physics of Nuclear Reactors

2.0 INTRODUCTION

Nuclear fission was discovered by Chadwick in 1939. This opened up the prospects of a new source of power utilizing the internal energy of the nucleus of an atom. The materials that were amenable to fission were uranium and thorium. In order to understand the characteristics of fission reactions, it is essential to have knowledge of the basics of atomic and nuclear physics. An atom comprises of a positively charged nucleus surrounded by negatively charged particles called electrons. Nucleus contains positively charged protons and neutrons that are electrically neutral. For a given element the number of protons is equal to the number of electrons, is referred to as the Atomic Number (Z). The total number of nucleons i.e. protons and neutrons in a nucleus is referred to as the mass number (A). The number of neutrons is A-Z. The masses of neutrons and protons are close to unity on the atomic mass scale. This chapter takes the subject of atom to the process of fission wherein huge heat energy is released. It also introduces the reader to the different terms commonly used in nuclear reactor physics.

2.1 ISOTOPES

Chemical property of an element is a function of the electrons which are orbiting around the nucleus. Atoms with same number of electrons/protons but with different number of neutrons will behave identically in a chemical reaction but exhibit marked differences in the nuclear characteristic. Elements having same atomic number but different mass numbers are referred to as Isotopes. Oxygen has three isotopes O^{16}, O^{17}, O^{18}. Commonly known oxygen is O^{16}, which has 8 protons and 8 neutrons, while O^{17} has 8 protons and 9 neutrons. Hydrogen has three isotopes H^1, H^2 and H^3. The second isotope is referred to as Deutrium (D) and the third one tritium (T). While ordinary water has the form H_2O, Heavy water has the form D_2O. The atomic mass unit (amu) is defined as exactly 1/16 of mass of O^{16} atom and is equal to $1.66*10^{-24}$ g. Mass of single proton is 1.007596 amu ($1.6725*10^{-24}$g), while mass of neutron is 1.008936 amu ($1.674*10^{-24}$) and that of electron is only 0.000549 amu ($9.11*10^{-28}$ g). Uranium is one of the most important elements as far as nuclear energy is concerned. It exists in nature in three isotopic forms with mass numbers 234, 235 and 238. The proportion of U_{234} is 0.0058, U_{235} is 0.711 and U238 is 99.283. The respective masses are 234.1140, 235.1175 and 238.1252 amu respectively. Thorium (Th_{232}) is another important element for nuclear energy but it has no isotopes.

2.2 BINDING ENERGY

Assessment of the mass of elements by mass spectrograph has shown that the actual mass is always less than the sum of the masses of protons and neutrons. The difference is referred to as mass defect and is related to the energy binding the particles in the nucleus.

$$\text{Mass defect } m = [Z(m_p + m_e) + (A - Z)m_n] - M$$

Hydrogen atom has 1 proton and 1 electron and the weight of 1 proton and 1 electron is the weight of hydrogen atom. Hence the above can be rewritten as

$$\text{Mass defect } m = [Zm_H + (A - Z)m_n] - M$$

m_H is 1.008145 and m_n is 1.008986 amu and mass defect can be evaluated for all atoms of different elements. Based on Einstein's special theory of relativity, the mass defect multiplied by c^2, where c is the velocity of light, is a measure of the energy that would be released when Z protons and A − Z neutrons are brought together to form a nucleus. If we are able to give the same amount of energy to a nucleus, it would be able to break the nucleus into protons and neutrons. The energy equivalent of mass defect is called the binding energy of the nucleus. Taking velocity of light as 2.998×10^{10} cm/s the energy E in ergs is obtained as

$$1E \text{ (erg)} = m(g) \times 8.99 \times 10^{20}$$

Converting E to MeV (1 MeV is 1.602×10^{12} ergs) and m to amu (1 amu = 1.66×10^{-24} g) we get

$$E \text{ (MeV)} = 931 \times m$$

Considering U235, the isotopic mass is 235.1175 and atomic number is 92. The binding energy is calculated as

$$BE/A = 931/235[(1.00814 \times 92) + (1.00898 \times 143) - 235.1175]$$
$$= 7.35 \text{ MeV per nucleon (neutron + proton)}$$

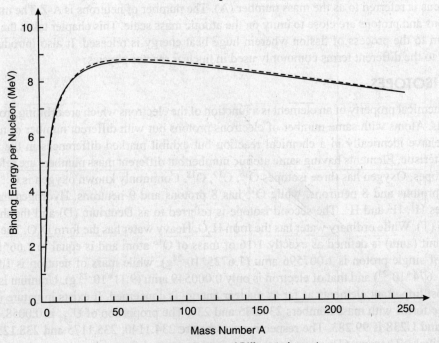

Fig 2.1: Binding Energy of Different elements

Basic Physics of Nuclear Reactors

Curve of BE per nucleon for elements of different mass numbers are given in Fig. 2.1. It can be seen that BE per nucleon is low for low mass numbers and it rises to nearly 8.7 MeV for iron (Fe of mass number 56). The BE per nucleon then falls gradually. Thus iron can be said to be the most stable element as it has the maximum BE.

From this curve we can note the following: Nuclides which are below Fe will have a tendency to become stable by means of fusion process to become a higher mass number element which is more stable. Nuclides which are above would have a tendency to break up by means of fission process to form nuclides with mass numbers close to Fe.

2.3 NUCLEAR STABILITY

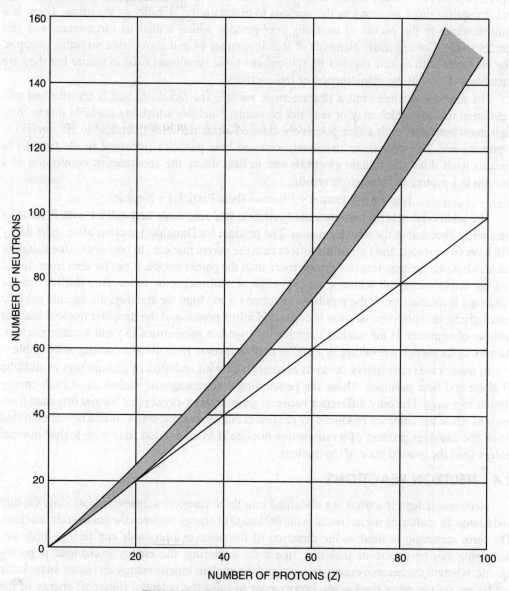

Fig 2.2: Neutron/proton ratio for stable elements

Fig. 2.2 shows a plot of the number of neutrons vs the number of protons in a stable nucleus. It can be seen that for stable nuclei of low mass numbers, the number of neutrons and protons are nearly equal. With increasing mass number, the nucleus is stable only with more neutrons than protons. For atomic numbers of 80 and above, the neutron/proton ratio is about 1.5.

The above characteristics can be accounted by considering the two types forces that exist between nucleons. First there are attractive forces of nearly equal magnitude between the nucleons i.e. protons attract other protons, neutrons attract neutrons and protons attract neutrons to about the same extent. These are intranuclear forces operative over short distances (10^{-15} m). In addition to the short range attractive forces, there are electrostatic repulsive forces between the protons that act over long distances. The total electrostatic forces between protons are proportional to the square of their number i.e. Z^2. For smaller mass number elements, the repulsive forces are small, but for larger mass number elements, the repulsive forces are large and it requires more neutrons in the nucleus to maintain the stability of the atom. There is a limit however to the excess of neutrons over protons which a nucleus can contain and still remain stable. Consequently elements of atomic number 84 and above have no stable isotopes. The elements with atomic number 84 (polonium) to 92 (uranium) exist in nature but they are unstable and exhibit the phenomenon of radioactivity.

The unstable nuclide emit a characteristic particle (or radiation) and is transformed into a different nuclide which may or may not be stable. Nuclides which are unstable due to their high mass numbers, emit either positively charged alpha particles (identical to He nuclei i.e. 2 protons and 2 neutrons), or negatively charged beta particles (identical to electrons). The nucleus itself does not contain electrons and in beta decay the spontaneous conversion of a neutron to a proton and a electron results.

$$\text{Neutron} \rightarrow \text{Proton} + \text{Electron (beta Particle)} + \text{Neutrino}$$

The additional neutral particle called neutrino has zero mass and carries with it some of the energy liberated in the transformation. The product (or Daughter) nucleus after alpha decay will have two protons and two neutrons less than the parent nucleus. In beta decay, the daughter nucleus has one neutron less one proton more than the parent nuclei. It can be seen from fig 2.2 that the stable nuclei fall within a narrow range of neutron/proton ratios. Any nuclide outside this range is radioactive. If the number of neutrons is too high for stability, the nuclide exhibits beta activity, in which one neutron is converted into a proton and the daughter nucleus has less number of neutrons. If the nuclide contains less number of neutrons to yield a stable nucleus, it emits alpha particles resulting in a higher neutron/proton ratio, thus becoming more stable.

In many cases radioactive decay is accompanied with emission of gamma rays in addition to alpha and beta particles. These are penetrating electromagnetic radiations of high energy similar to x-rays. The only difference between gamma and x-rays is that former originate from nucleus while the latter are produced by processes outside the nucleus. Gamma rays are emitted when the daughter product of a radioactive nuclide is in an excited state with higher internal energy than the ground state of the nucleus.

2.4 NEUTRON REACTIONS

Neutron-nucleus reactions are classified into three categories, namely, scattering, capture and fission. In scattering the net result is the exchange of energy between the neutron and nucleus. The term scattering is used as the direction of the neutron afterwards can be any direction. Scattering can be elastic or inelastic. In elastic scattering, the energy exchanged is purely kinetic, wherein the neutron energy reduces and the nucleus kinetic energy increases. In inelastic scattering, on the other hand some energy goes to raise the potential (internal) energy of the nucleus. When first liberated in fission free neutrons possess high energy (~MeV) and are called fast neutrons. They lose the kinetic energy as they undergo collisions and become slow neutrons

(~ eV). Generally the first stage in a neutron-nucleus interaction is the absorption of the neutron by the nucleus to form a compound nucleus in an excited state. In inelastic scattering, the compound nucleus immediately expels a neutron of lower energy. If the compound nucleus emits its excess energy as gamma radiation, it is referred to as radiative capture. It is called a (n, γ) reaction. Example of a neutron reactions important to nuclear reactors is given below.

The capture of neutron U^{238} results in a (n, γ) reaction.

$$_{92}U^{238} + {_0}n^1 \rightarrow {_{92}}U^{239} + \gamma$$

The resulting Uranium 239 is radioactive and decays with emission of a beta particle

$$_{92}U^{239} \rightarrow {_{-1}}\beta^0 + {_{93}}Np^{239}$$

Neptunium 239 is also radioactive and decays as given below.

$$_{93}Np^{239} \rightarrow {_{-1}}\beta^0 + {_{94}}Pu^{239}$$

2.5 RADIOACTIVE DECAY LAW

The fundamental law of radioactive decay is that for a specified type of nucleus, the probability of a given nucleus decaying in a unit of time is a constant, i.e. independent of external conditions or the number of nuclei present. Stated another way, it is an experimental fact that the number of disintegrations occurring in unit time is proportional to the number of radioactive atoms present. From either approach the mathematical expression for the rate of decay is

$$dN/dt = -\lambda N$$

where N is the number of radioactive atoms at a given time λ is the "decay constant," which is a measure of the decay probability.

If at any arbitrary zero time, N_0 radioactive atoms are present, it is found by integration of above equation that—

$$N = N_0 e^{-\lambda t} \text{ (Fig 2.3)}$$

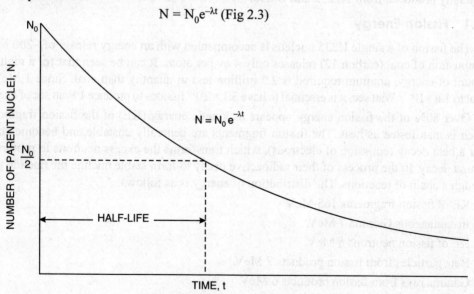

Fig. 2.3: Exponential decay of radioactivity

The "mean life" or "average life," designated t_m, of a radioactive species is related to the decay constant by the expression:

$$t_m = 1/\lambda$$

It is customary to express the decay of a radioactive species in terms of the "half life" $t_{1/2}$ It is defined as the time required for the number of the radioactive atoms present to decay to half of the initial value.

By substitution of N equals $N_0/2$ it is found that
$$t_{1/2} = 0.6931/\lambda = 0.6931\, t_m$$

2.6 UNITS OF RADIOACTIVITY

The "curie" expresses radioactive intensity in terms of the number of atoms which disintegrate per unit time. The curie is defined as that quantity of any radioactive material which gives 3.7×10^{10} disintegrations per sec. The "roentgen" is a unit which measures the amount of ionization and thus the amount of energy, produced by electromagnetic radiation, X- or gamma rays. The formal definition is "The roentgen shall be the quantity of X- or gamma radiation such that the associated corpuscular emission per 0.001293 gm of air produces, in air, ions carrying one electrostatic unit of quantity of either sign." The value of the energy absorbed per roentgen is 0.108 ergs per cubic centimeter of air or 83.8 ergs per gram of air. The "rad " is a unit of deposition of energy by ionization equal to 100 ergs per gram of absorbing material. The rad is preferred because its application is more straightforward than that of the roentgen in many situations.

2.7 FISSION

When a neutron is able to transfer an amount of energy equal to or more than the BE of that element, the nucleus break up into two lighter nuclei. The process of breaking of an atom by neutron is called fission and the lighter nuclei formed are referred to as fission products or fragments. Three nuclides i.e. U233, U 235 and Pu 239 are fissionable by neutrons of all energies. Of these only U 235 occurs in nature (0.71% of natural uranium) and the other two are artificially produced from Th 232 and U 238.

2.7.1 Fission Energy

The fission of a single U235 nucleus is accompanied with an energy release of ~200 MeV. Combustion of coal (carbon 12) releases only 4 ev per atom. It can be seen that for a required amount of energy, uranium required is 2.5 million less in quantity than coal. Since 1 MeV is equal to 1.6×10^{-12} Watt sec, it is essential to have 3.1×10^{10} fissions to produce 1 Watt sec of energy.

Over 80% of the fission energy appears as kinetic energy (KE) of the fission fragments, which is manifested as heat. The fission fragments are generally unstable and become stable after a beta decay (emission of electrons), which transforms the excess neutrons to protons or gamma decay. In the process of their radioactive decay to form stable nuclide the nuclide goes through a chain of reactions. The distribution of energy is as follows:

KE of fission fragments 165 MeV.

Instantaneous Gamma 7 MeV.

KE of fission neutrons 5 MeV.

Beta particles from fission products 7 MeV.

Gamma rays from fission products 6 MeV.

Neutrinos 10 MeV.

2.7.2 Critical Mass

In a fission reaction two to three neutrons are released on an average. Out of these some may be captured in non fissile material and some neutrons may leak out without encountering any fissile atom and rest only available for fission. It can happen that the net number of neutrons available to continue with fission is much less than one. In such a case the fission reaction chain cannot be sustained. The neutron fraction lost due to leakage from the boundaries can be reduced by increasing the number of fissile atoms or the mass of fuel. The minimum amount of such material that is capable of sustaining the fission reaction for a given geometry is referred to as the critical mass. It may be noted that the critical mass is very much a function of the geometry in which the fissile materials are arranged.

2.7.3 GABON- The Natural Reactor

In May 1972 at the Pierrlette uranium enrichment facility in France, routine mass spectrometry comparing UF6 samples from the Oklo mine, located in Gabon, Central Africa showed a discrepancy in the amount of the 235 U isotope. Normally the concentration is 0.720%; these samples had only 0.717% – a significant difference. This discrepancy required explanation, as all uranium handling facilities must meticulously account for all fissionable isotopes to assure that none are diverted for weapons purposes. Thus the French Commissariat a l' energie atomique (CEA) began an investigation. A series of measurements of the relative abundances of the two most significant isotopes of the uranium mined at Oklo showed anomalous results compared to those obtained for uranium from other mines. Further investigations into this uranium deposit discovered uranium ore with a 235U concentration as low as 0.440%. This loss in 235U is exactly what happens in a nuclear reactor. A possible explanation therefore was that the uranium ore had operated as a natural fission reactor. Other observations led to the same conclusion, and on September 25, 1972, the CEA announced their finding that self-sustaining nuclear chain reactions had occurred on Earth about 2 billion years ago. Later, other natural nuclear fission reactors were discovered in the region.

2.7.4 Liquid Drop Model of Fission

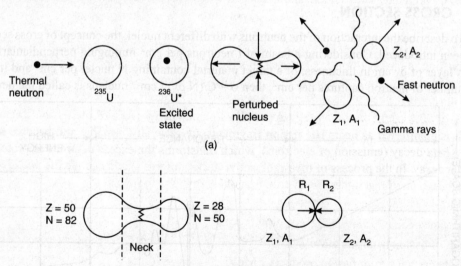

Fig. 2.4: Liquid drop model of fission of U235

Some of the characteristics of the fission process can be obtained by the use of liquid drop model. A nucleus is akin to a drop of liquid interacting with other nuclei or liquid drops in the vicinity. The Binding energy of the nucleus is proportional to the mass number. The particles on the surface have less number of neighbors than the interior ones. Consider a liquid drop to which a force is applied. It passes through different stages as shown in the figure 2.4. From a spherical shape, it becomes elliptical. Though volume has remained same the surface area has increased. With sufficient force the drop would get into dumbbell shape, with more surface area. With larger surface the surface energy would exceed the internal cohesive energy of the liquid drop. At this stage the drop would split into two droplets. The situation in case of fission is similar. A target nucleus absorbs a neutron and the resulting excitation energy is equivalent to the binding energy of the neutron plus the kinetic energy of the neutron before capture. As a result of the excess energy, the compound nucleus undergoes deformation. While the attractive forces (internal) compel the nucleus to return to its original shape, the excess energy results in more deformation. Depending on the excess energy the compound nucleus would split into two fission fragments. The minimum excess energy required to break is referred to as critical energy. Values of critical energy and neutron binding energy for a number of nuclide are given in table 2.1.

Table 2.1

Target Nucleus	Critical Energy MeV	Neutron Binding Energy MeV
Th232	5.9	5.1
U238	5.9	4.8
U235	5.8	6.4
U233	5.5	6.7
Pu239	5.5	6.4

From the above it can be seen that the neutron binding energy for the fissile nuclides U-235, U-233 and Pu-239 exceeds the critical energy needed for fission. Hence capture of a neutron of zero energy would be sufficient to cause fission in these nuclei. If the captured neutrons have more than 0.8 MeV of energy, it can cause fission in Th232. For fission in U238 the neutron must have energy greater than 1.1 MeV.

2.8 CROSS SECTION

To describe the interaction of the neutrons with different nuclei, the concept of cross section has been introduced. Considering a beam of I neutrons per cm^2 impinging perpendicularly on a thin layer of δx cm in thickness, of a target material containing N nuclei per cm^3 and if C is the number of neutron captures per cm^2, then $\sigma = C/(N \delta x) I$ $cm^2/$ nucleus is called the neutron

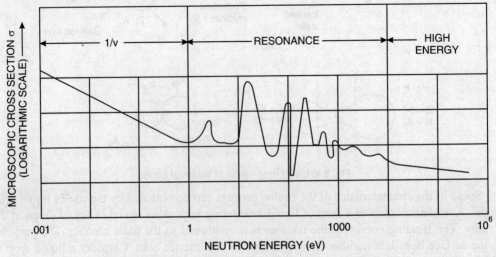

Fig 2.5: Variation of fission cross section with neutron energy

Basic Physics of Nuclear Reactors

cross section. Since these cross sections are in the range of 10^{-12} to 10^{-24} cm^2 per nucleus, it is a general practice to express them in barns (1 barn = 10^{-24} cm^2/nucleus). Every nuclide has a cross section for fission, for absorption or scattering the neutron through elastic or inelastic collisions. Fission cross sections are a function of the energy carried by the neutron. In general, neutrons with energies less than 1 eV are called as slow or thermal neutrons, while neutrons with energies greater than 1 MeV are called fast neutrons.

Table 2.2: Thermal neutron Cross Sections for different nuclei

Nucleus	Fission Cross Section (barns)	Capture Cross section (barns)	Total Absorption Cross Section (barns)
U 233	527	54	581
U 235	577	100	683
Pu 239	742	287	1029
U 238	–	2.71	2.71
U natural	4.2	3.5	7.7

The neutrons with energies between 1 eV to 1 MeV are termed as intermediate/epithermal neutrons. In the intermediate region, there are energies at which the cross sections are very high and this is also referred to as resonance region to indicate resonance effects (Fig 2.5). As the energy increases the probabilities of interaction reduces and hence cross sections for fission or absorption reduce with neutron energy. For neutrons travelling with a velocity of 2200 m/s, which are in our interest range the neutron cross sections are given in table 2.2 for different nuclides. It can be seen that in the case of U-235 the fission is more efficient than Pu 239. In the former 84% of the neutrons absorbed cause fission, while for the latter only 72% of neutrons absorbed cause fission. For the case of N number of fissile nuclei per cm^3, σ_f cm^2 per nucleus, the fission cross section σ_f and φ, the number of neutrons per cm^2 per second, the fission rate is given by

$$F = N * \sigma_f^* \, \varphi \text{ fissions/cm}^3/\text{s}$$

Considering a reactor with a fuel volume of V cm^3 and the fact that $3.1*10^{10}$ fissions per second are essential to produce 1 watt of power, Power P is given by

$$P = V * N * \sigma_f^* \, \varphi / 3.1*10^{10} \text{ Watts}$$

In the above equation, V* N represents the total number of fissile nuclei and upon division by Avogadro's number and multiplication by atomic weight, the mass of fissile material (U-235) can be obtained as

$$G = 235 * V * N / 0.602 * 10^{24} \text{ gm}$$

2.9 PROMPT AND DELAYED NEUTRONS

Neutrons that are released immediately after fission are termed as prompt neutrons. A small fraction (0.65%) of neutrons is emitted from the fission fragments after they undergo beta decay. Such neutrons are referred to as delayed neutrons and the delay time is the time required for beta decay. Fission products that lead to delayed neutrons are termed precursors.

The time required for a prompt neutron to be born, to come down in energy to thermal neutron level and cause a fission is on the order of 10^{-4} s. This is too rapid for human or machine control. Delayed neutrons have an average lifetime of 12.2 s. The effective fraction of neutrons that are delayed at thermal energies in a typical light-water reactor is 0.0065. This quantity is denoted by the Greek letter β and is called Beta. The lifetime of an 'average' neutron is therefore:

(number of prompt n's) (prompt lifetime) + (number of delayed n's)(delayed lifetime)/ total number of neutrons

Assuming that a reactor has 100,000 neutrons present, the average lifetime of neutron is
(99350)*(0.0001) + (650) *(12.2)/ 100,000 = 0.079 s

Thus delayed neutrons lengthen the average neutron llifetime and result in a controllable reactor.

2.10 NEUTRON LIFE CYCLE (FIG 2.6)

A clear understanding of the neutron life cycle is essential for the design of a nuclear reactor. In any nuclear fission, fast neutrons will be produced. Some of these fast neutrons will cause fission and some will be slowed down to produce a thermal fission. These fast neutrons will slow down or will be absorbed in the reactor through multiple steps. These steps are referred to as the neutron life cycle. In this section, the different factors or parameters of the neutron life cycle will be discussed. The neutron life cycle quantifies the possible events that might occur as a neutron is thermalised. The term thermalised is used to refer to bringing the neutron energy to thermal energy level.

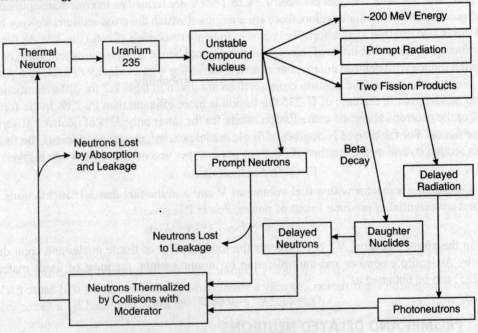

Fig 2.6: Neutron Life Cycle

2.11 INFINITE MULTIPLICATION FACTOR (K_∞)

The probability that all neutrons produced by a fission reaction will cause another fission is not always true, because some of these neutrons will leak out of the reactor or will be absorbed in the reactor structure or in the fuel itself but without causing a fission. To sustain the chain reaction, for each fissioned nucleus, there should be at least one neutron that causes another fission. The multiplication factor is the term briefly describing this condition.

The multiplication factor is determined by finding out how many neutrons are absorbed or leaked out from the reactor. This will determine whether the number of new generation neutrons

Basic Physics of Nuclear Reactors

will be less, more or the same amount as the previous generation. Depending on the size of the reactor core, neutron leakage from the reactor will be more or less. If the reactor size is large, neutron leakage will be small and vice versa. Also if the reactor is considered infinitely large, there will be no neutron leakage. In this situation the infinite multiplication factor, k_∞, is the measure of the increase or decrease in neutron flux and is calculated by the ratio of the neutrons produced by fission in one generation to the number of neutrons lost through absorption in the preceding generation. This can be expressed mathematically as shown below.

$$k_\infty = \frac{\text{neutron production from fission in one generation}}{\text{neutron absorption in the preceding generation}}$$

2.11.1 Four Factor formula

During the neutron life cycle in an infinitely large reactor, the fast neutrons produced by fission will go through multiple steps. Some of these steps will decrease the number of these neutrons and some will increase it or even produce a second generation by causing another fission. Independent of the size and shape of the reactor, there are four factors that describe the neutron life cycle. These four factors will represent the infinite multiplication factor as shown in the equation below.

$$k_\infty = \varepsilon p f \eta$$

where:

ε = fast fission factor
p = resonance escape probability
f = thermal utilization factor
η = reproduction factor

Each of these four factors represents a process that will increase or decrease the initial number of neutrons produced by fission in a generation. In the following sections, the four factors are explained.

2.11.1.1 Fast Fission Factor (ε)

Firstly, the neutrons of a generation will go through a fast fission. This means that these neutrons are in the fast energy region. After this process, the number of fast neutrons in the reactor will increase. Depending on the type of the reactor (either thermal or fast), if the reactor is a thermal reactor, the fuel will be mainly Uranium-235 that has a small probability (referred as a cross section) for fast fission. So only a small number of fast neutrons will cause fission. In any situation, the number of fast neutrons in a generation is increased by a factor called the *fast fission factor*. The fast fission factor (ε) is defined as the ratio of the number of fast neutrons produced by all fissions to the number of fast neutrons produced by thermal fissions. The mathematical expression of this ratio is shown below.

$$\varepsilon = \frac{\text{number of fast neutrons produced by all fissions}}{\text{number of fast neutrons produced by thermal fissions}}$$

In order for a fast fission to occur, a neutron must hit a fuel nucleus while this neutron is still in its fast energy region. So the fast fission factor will be affected by the concentrations and arrangement of the fuel and the moderator. For a homogenous reactor system where all the fuel atoms are surrounded by moderator atoms, the fast fission factor equals one, but for a heterogeneous reactor where all the fuel atoms are packed closely together in elements (rods, pellets, or pins), fast neutrons produced from a fission of one fuel atom have a high probability for causing fissions in other fuel atoms while these neutrons are still in their fast energy region. A typical value of the fast fission factor in most heterogeneous reactors is about "1.03".

2.11.1.2 Resonance Escape Probability (p)

After the fast fission process, the neutrons continue to spread throughout the reactor. These neutrons will slow down and lose some of their energy as they collide with nuclei of fuel and non-fuel materials and moderator in the reactor.

While slowing down, these neutrons will go through the resonance region of Uranium-238 (from 6eV to 200eV) and there is a chance that some of these neutrons will be captured in this region. The probability that a neutron will not be absorbed in the resonance region is called the *resonance escape probability* (*p*). The resonance escape probability is defined as the ratio of the number of neutrons that reach thermal energies to the number of fast neutrons that start to slow down. The mathematical expression of this ratio is shown below.

$$p = \frac{\text{number of neutrons that reach thermal energy}}{\text{number of fast neutrons that start to slow down}}$$

The fuel with moderator arrangement and the enrichment of Uranium-235 (in light water reactor) are playing a main role in determining the value of the resonance escape probability. This value is higher in a heterogeneous reactor than it is in a homogeneous reactor because in the former one the neutron slows down in the moderator where there are no atoms of Uranium-238 present, but in the later one the neutron slows down in the fuel where there are a lot of Uranium 238 atoms.

After the fast fission and the resonance escape processes, the left over neutrons will be equal to the ratio of fast neutrons that survive slowing down or thermalization to the number of fast neutrons initially started the generation.

2.11.1.3 Thermal utilization factor (f)

After the slowing down process, the neutrons continue to spread throughout the reactor and are absorbed either by the fuel or any other materials in the reactor. The *thermal utilization factor* (*f*) is a measure of how well thermal neutrons are absorbed by the fuel. The thermal utilization factor is defined as the ratio of the number of thermal neutrons absorbed in the fuel to the number of thermal neutrons absorbed in any reactor material. The mathematical expression of this ratio is shown below.

$$f = \frac{\text{number of thermal neutrons absorbed in the fuel}}{\text{number of thermal neutrons absorbed in all reactor materials}}$$

This factor will always be less than one because a part of the thermal neutrons will be absorbed in non-fuel materials.

2.11.1.4 Reproduction Factor (η)

Mainly, most of the neutrons absorbed in the fuel cause fission, but that is not always true because part of these neutrons absorbed in the fuel may not cause fission. The *reproduction factor* (η) is defined as the ratio of the number of fast neutrons produces by thermal fission to the number of thermal neutrons absorbed in the fuel. The reproduction factor is shown in the equation below.

$$\eta = \frac{\text{number of fast neutrons produced by thermal fission}}{\text{number of thermal neutrons absorbed in the fuel}}$$

Basic Physics of Nuclear Reactors

2.12 EFFECTIVE MULTIPLICATION FACTOR (k_{eff})

To describe the neutron life cycle in a finite reactor effectively, it is a requirement to account for neutrons that leak out from reactor. The *effective multiplication factor* (k_{eff}) takes this neutron leakage into account. The effective multiplication factor is defined as the ratio of the neutrons produced by fission in one generation to the number of neutrons lost through absorption and leakage in the preceding generation. This factor may be expressed mathematically as shown below.

$$k_{eff} = \frac{\text{neutron production from fission in one generation}}{\text{neutron absorption in the preceding generation} + \text{neutron leakage in the preceding generation}}$$

So, in the condition of self-sustaining chain reaction of fissions, the effective multiplication factor should equal one. This condition is called the *critical condition* and in this condition the neutron population is neither increasing nor decreasing. However if the neutron production is greater than the neutron absorption and leakage, the reactor will be in a *supercritical condition* and k_{eff} is greater than one. Alternatively, if the neutron production is less than the absorption and leakage, the reactor will be in a *subcritical condition* and k_{eff} is less than one.

For a finite reactor, this factor may be expressed mathematically in terms of the infinite multiplication factor and two additional factors which account for neutron leakage as shown below.

$$k_{eff} = k_\infty L_f L_t$$

2.12.1 Fast Neutron Non-Leakage Probability (L_f)

In a finite reactor, some of the fast neutrons leak out of the boundaries of the reactor core before they begin the slowing down process or before they get thermalized. The *fast non-leakage probability* (L_f) is defined as the ratio of the number of fast neutrons that do not leak from the reactor core to the number of fast neutrons produced by all fissions. This may be expressed mathematically as follows.

$$L_f = \frac{\text{number of fast neutrons that do not leak from reactor}}{\text{number of fast neutrons produced by all fissions}}$$

2.12.2 Thermal Neutron Non-Leakage Probability (L_t)

Also in a finite reactor, some of the neutrons leak out of the boundaries of the reactor core after they slowed down or after they got thermalized. The *thermal non-leakage probability* (L_t) is defined as the ratio of the number of thermal neutrons that do not leak from the reactor core to the number of neutrons that reach thermal energies. The thermal non-leakage probability is expressed mathematically by the following.

$$L_t = \frac{\text{number of thermal neutrons that do not leak from reactor}}{\text{number of neutrons that reach thermal energies}}$$

The *total non-leakage probability* (L_T) that presents the fraction of all neutrons that don't leak out of the reactor core could be calculated by combing the two probabilities (L_f & L_t). This term can be expressed as:

$$L_T = L_f L_t$$

2.12.3 Six Factor Formula

To precisely follow the neutron life cycle in a finite reactor, the last two factors should be added to the four factors formula to get the six factors formula and it is expressed as the effective multiplication factor (k_{eff}). This may be expressed mathematically as follows.

$$k_{eff} = \varepsilon L_f p L_t f \eta$$

Using this six factor formula, any neutron produced by fission could be traced from its production to the initiation of subsequent fissions. Figure 2.7 illustrates a neutron life cycle with nominal values provided for each of the six factors.

2.13 NEUTRON MODERATION

It may be noted that the neutrons are born at energies that are well above both the thermal (0.025 eV) and epithermal/intermediate (keV range) regions. Such neutrons both prompt and delayed, are termed "fast". The probability of a fast neutron being absorbed is very small, as the absorption cross sections are low. However, it is substantial (~582 barns) for U-235 at thermal energies. So, for a fission chain reaction to be sustained, it is essential that the fission neutrons be slowed down or thermalized. This process is called neutron moderation.

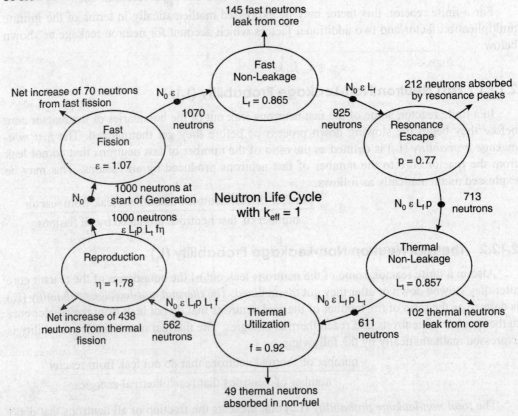

Fig 2.7: Example of neutron life cycle

2.14 NEUTRON SLOWING DOWN

The fast neutrons are slowed down by collisions. From energy and momentum conservation, it can be shown that the maximum energy transfer between two particles in a head on collision is:

Basic Physics of Nuclear Reactors

$$Q_{max} = \frac{4mM}{(m+M)^2} E$$

Where

Q_{max} is the energy transferred
m is the mass of the incident particle
M is the mass of the moderator material, and
E is the energy of the incident particle.

This relation indicates that a head-on collision between a neutron and another particle of equal mass (e.g., a proton) will result in a complete transfer of energy from the neutron to the proton. In contrast, a collision between a neutron and a particle of large mass (i.e., $m \ll M$), will result in the transfer of almost no energy. Thus, moderators of low atomic mass are desired.

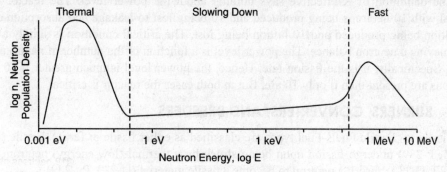

Fast: 1 – 10 MeV, Zone of Neutron Birth From Fission

Slowing Down: 1 MeV – 1 eV, Zone of Neutron Collisions With Moderator Nuclei, and Kinetic Energy Exchange to Target Nuclear; Zone of Resonance Absorption

Thermal: < 1 eV, Zone of Neutron Elimination (via Absorption or Escape); Zone of Kinetic Energy Exchange To, From Neutron and Target Nucleus; Rate Depends Upon Temperature of Target Medium

Fig 2.8: Neutron Slowing Down

2.15 NEUTRON MODERATORS

Possible moderators together with their advantages/disadvantages are:

(a) *Light-Water:* Best choice for efficient scattering. Drawback is that its neutron absorption cross-section is significant.

(b) *Heavy Water:* Very little absorption of neutrons. The scattering cross-section is less effective than that of light water. Also, the absorption reaction, although minor, yields tritium which is a hazard to those doing maintenance on the reactor.

(c) *Beryllium:* Low absorption, reasonable scatter cross-section. Beryllium dust, if inhaled, can cause serious lung disease that is not curable.

(d) *Graphite:* Low absorption, reasonable scatter cross-section. Drawbacks are that it is combustible. Fast neutrons can cause separation of the graphite atoms which are layered. This stored energy accumulates in the graphite and can cause combustion unless annealed out. This is referred to as Wigner Effect.

A comparison of the properties of the four materials is given in table 2.3 below.

Table 2.3: Moderating capability of materials

Material	Scattering Cross Section (barns)	Absorption Cross Section (barns)	Density (g/cm^3)	Atomic Weight
H_2O	103	0.66	1.00	18
D_2O	13.6	<0.00134	1.105	20
Be	6.14	0.0092	1.85	9
G	4.75	0.0034	1.60	12

2.16 CRITICALITY AND REACTOR POWER

The definition of K-effective says nothing about the power level. The reactor can be critical with 10 neutrons being produced and 10 being lost to leakage and absorption or with 10 billion being produced and 10 billion being lost. The critical condition is simply a matter of achieving a neutron balance. The power level is a function of the number of neutrons in the cycle. Specifically, it is the fission rate. Hence, the power level is much greater if 10 billion neutrons are present than if only 10 are. But in both cases the reactor is critical.

2.17 BURNERS, CONVERTERS, AND BREEDERS

Fissile vs. Fertile Fuels-Fuel types are classified as either fissile or fertile. Fissile (U-233, U-235, P-239) undergo fission upon being struck by a thermal (low energy) neutron. Fertile (Th-232, U-238) absorb a neutron to become a fissile material (U-233, Pu-239).

Under what conditions can we design a reactor to produce fissile isotopes from fertile ones? Three possible fuel cycles are:

Breeder: More than one fissile atom is produced for each one that is consumed.

Converter: Converts fertile material to fissile but does not breed.

Burner: Neither converts nor breeds.

The symbol ν denotes the number of neutrons produced per fission. For thermal fission of U-235, its value is 2.5. This quantity by itself is insufficient to determine if conversion or breeding is possible because it only states the number of neutrons emitted once fission occurs. We need to know the number of neutrons emitted once a neutron is absorbed by a fissile nucleus. We denote that parameter as η and define it as

$$\eta = \nu \frac{\sigma_f}{\sigma_a} = \nu \frac{\sigma_f}{\sigma_\gamma + \sigma_f}$$

Where σ_f is the microscopic fission cross-section

σ_a is the microscopic absorption cross-section, and

σ_γ is the microscopic capture cross-section.

For even a burner to operate, η must be greater than 1 because some neutrons are lost to processes other than absorption in U-235. Some are absorbed in the core structure, some diffuse out into the shield, and others are absorbed by the steel, etc., that forms the reactor's structure. For breeding, we need an additional neutron from each fissile absorption in order to convert fertile nuclei to fissile ones. Thus, η must exceed 2. η is a function of the energy of the incident neutron. Fig 2.9 shows the values of η for U-233, U-235 and Pu-239.

Basic Physics of Nuclear Reactors

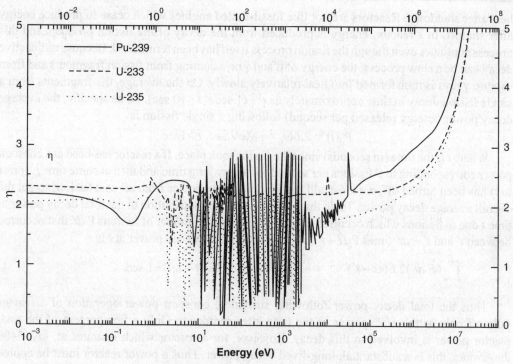

Fig 2.9: η values for different fissile nuclei

The value of η for the three fissile isotopes are given below:

Table 2.4: η Values for Fissile Nuclide

Neutron Energy	U_{233}	U_{235}	Pu_{239}
< 0.025 eV	2.29	2.07	2.14
> 100 KeV	2.31	2.10	2.45

η = 2.29 is sufficiently above 2.0 so that a U-233/Th-232 core would produce more U-233 then it consumes. The values of η for both U-235 and Pu-239 are also above 2.0, but not so much above it that breeding is not practical. So, at thermal energies, only the U-233/Th-232 cycle is possible for breeding.

In the intermediate or epithermal neutron energy levels (1eV – 100 keV) for U-235 and 10-20 keV for Pu-239, the value of η is below 1.0. For U-233, the value of η might be able to sustain breeding, but this has not been established.

In the fast neutron energy levels (>100 keV) the value of η is sufficiently above 2.0 for all three fuels to function as a breeder.

There is no fertile nuclide that transmutes to U-235 upon neutron absorption. So, once supplies of U-235 have been consumed, it is gone. We can have a U-233/Th-232 cycle or a Pu-239/U-238 cycle. For U-235, the cycle would most likely involve Pu-239 and U-238 with the U-235 present only initially.

2.18 DECAY HEAT

When nuclear energy was first proposed, its advocates are claimed to have said that it "would be too cheap to meter." And perhaps it would have been, except for the production of heat after shutdown. Reactors are not like fossil-fueled engines which cease to produce energy after shutoff. In a reactor, energy is produced from the decay of the fission products and this process continues even though the fission process itself has been terminated. Because radioactive decay can be a slow process, the energy of β and γ rays coming from fission fragments and from capture γ rays is transformed into heat relatively slowly. On the average, the fragments from a single fission decay in time approximately as $t^{1/2}$ (1 sec $\leq t \leq$ 10 sec). To be specific, the average decay power (energy released per second) following a single fission is

$$P_d(t) = 2.66 t_0^{-1.2} \text{ MeV/sec}, t > 1 \text{sec}$$

Where t is the time (in seconds) since the fission took place. If a reactor has been at a constant power corresponding to F fissions per second for a very long time and then at some time t_0 (prior to t) has been turned off, we must add the average decay power from all past fissions to get the overall average decay power. To do this, we note that the contribution to the total decay power at time t due to fissions which occurred in time dt' equals the number of fissions F dt' that occurred between t' and $t' + dt'$ times $P_d(t - t')$. As a result the total decay power at t is

$$\int_{-\infty}^{t_0} (F\, dt') 2.66 (t - t')^{-1.2} = \frac{2.66}{0.2} F(t - t_0)^{-0.2}, (t - t_0) > 1 \text{ sec.}$$

Thus the total decay power following sustained, constant power operation of a reactor falls off only as the $1/5^{\text{th}}$ power of the time after shutdown. Only a few percent of the total reactor power is involved in this decay. However, for a reactor which operates at, say, 2400 megawatts, this is a substantial, long-lived decay power. Thus a power reactor must be cooled after shutdown.

The energy produced after reactor shutdown is called decay heat. If P_0 is power before shutdown and P is the decay heat (both in same units). Then,

$$\frac{P}{P_0} = 6.1 \times 10^{-3} [(\tau - t_0)^{-0.2} - \tau^{-0.2}]$$

Where $(\tau - t_0)$ is the time since shutdown in days (i.e., the cooling period) and τ is the time since startup.

2.19 REACTIVITY

Reactivity is a measure of the departure of a reactor from criticality. The reactivity is related to the value of k_{eff}. Reactivity is a useful concept to predict how the neutron population of a reactor will change over time. If there are N_o neutrons in the preceding generation, then there are $N_o * k_{eff}$ neutrons in the present generation. The numerical change in neutron population is $(N_o k_{eff} - N_o)$. The gain or loss in $(N_o k_{eff} - N_o)$ neutron population, expressed as a fraction of the present generation $(N_o k_{eff})$, is shown below.

$$\rho = (N_o k_{eff} - N_o) / (N_o k_{eff})$$

This relationship represents the fractional change in neutron population per generation and is referred to as *reactivity* (ρ). Cancelling out the term N_o from the numerator and denominator, the reactivity is determined as shown in the equation below.

$$\rho = (k_{eff} - 1)/k_{eff}$$

It may be seen that ρ may be positive, zero, or negative, depending upon the value of k_{eff}. The larger the absolute value of reactivity in the reactor core, the further the reactor is from criticality. It may be convenient to think of reactivity as a measure of a reactor's departure from criticality. Reactivity is a dimensionless number. It does not have dimensions of time, length, mass, or any combination of these dimensions. It is simply a ratio of two quantities that are dimensionless. The value of reactivity is often a small decimal value. In order to make this value easier to express, artificial units are defined. By definition, the value for reactivity that results directly from the calculation of $\Delta k/k$. Alternative units for reactivity are $\%\Delta k/k$ and pcm (percent milli).

$$1 \text{ pcm} = 10^{-5} * \Delta k/k$$

2.19.1 Feedback Reactivity

The amount of reactivity (ρ) in a reactor core determines the neutron population, and consequently the reactor power, at any given time. The reactivity can be affected by many factors (for example, fuel depletion, temperature, pressure, or neutron poisons/ absorbers). To quantify the effect that a variation in parameter (that is, increase in temperature, control rod insertion, increase in neutron poison) will have on the reactivity of the core, *reactivity coefficients* are used. Reactivity coefficients are the amount that the reactivity will change for a given change in the parameter. For instance, an increase in moderator temperature will cause a decrease in the reactivity of the core. The amount of reactivity change per degree change in the moderator temperature is the moderator temperature coefficient. Typical units for the moderator temperature coefficient are pcm/°C. Reactivity coefficients are generally symbolized by α_x, where x represents some variable reactor parameter that affects reactivity. The definition of a reactivity coefficient in equation format is shown below.

$$\alpha_x = \Delta\rho/\Delta x$$

If the parameter x increases and positive reactivity is added, then α_x is positive. If the parameter x increases and negative reactivity is added, then α_x is negative.

Reactivity change can be determined by multiplying the change in the parameter by the average value of the reactivity coefficient for that parameter.

$$\Delta\rho = \alpha_x * \Delta x$$

Changes in the physical properties of the materials in the reactor will result in changes in the reactivity.

2.19.1.1 Moderator Temperature Coefficient

The change in reactivity per degree change in temperature is called the *temperature coefficient of reactivity*. Because different materials in the reactor have different reactivity changes with temperature and the various materials are at different temperatures during reactor operation, several different temperature coefficients are used. Usually, the two dominant temperature coefficients are the moderator temperature coefficient and the fuel temperature coefficient.

The change in reactivity per degree change in moderator temperature is called the *moderator temperature coefficient* of reactivity. The magnitude and sign (+ or −) of the moderator temperature coefficient is primarily a function of the moderator-to-fuel ratio. If a reactor is under moderated, it will have a negative moderator temperature coefficient. If a reactor is over moderated, it will have a positive moderator temperature coefficient. A negative moderator temperature coefficient is desirable because of its self-regulating effect. For example, an increase in reactivity causes the reactor to produce more power. This raises the temperature of the coolant/moderator and adds negative reactivity, which slows down the power rise.

2.19.1.2 Fuel Temperature Coefficient

Another temperature coefficient of reactivity, the fuel temperature coefficient, has a greater effect than the moderator temperature coefficient for some reactors. The *fuel temperature coefficient* is the change in reactivity per degree change in fuel temperature. This coefficient is also called the "prompt" temperature coefficient because an increase in reactor power causes an immediate change in fuel temperature. A negative fuel temperature coefficient is generally considered to be even more important than a negative moderator temperature coefficient because fuel temperature immediately increases following an increase in reactor power. The time for heat to be transferred to the moderator is measured in seconds. In the event of a large positive reactivity insertion, the moderator temperature cannot affect the power rise for several seconds, whereas the fuel temperature coefficient starts adding negative reactivity immediately.

Another name applied to the fuel temperature coefficient of reactivity is the fuel Doppler reactivity coefficient. This name is applied because in typical low enrichment, light water moderated, thermal reactors the fuel temperature coefficient of reactivity is negative and is the result of the doppler effect, also called doppler broadening. The phenomenon of the Doppler effect is caused by an apparent broadening of the resonances due to thermal motion of nuclei as illustrated in Figure 2.10.

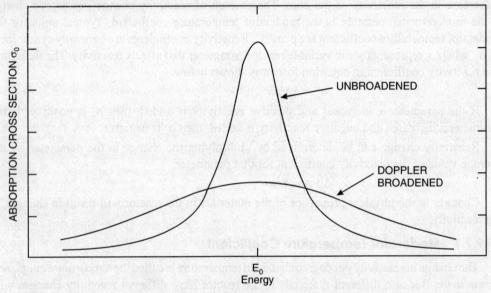

Fig. 2.10: Doppler Broadening of Cross Sections

Stationary nuclei absorb only neutrons of energy E_o. If the nucleus is moving away from the neutron, the velocity (and energy) of the neutron must be greater than E_o to undergo resonance absorption. Likewise, if the nucleus is moving toward the neutron, the neutron needs less energy than to be absorbed. Raising the temperature causes the nuclei to vibrate more rapidly within their lattice structures, effectively broadening the energy range of neutrons that may be resonantly absorbed in the fuel. Two nuclides present in large amounts in the fuel of some reactors with large resonant peaks that dominate the doppler effect are uranium-238 and plutonium-240.

2.19.1.3 Pressure Coefficient

The reactivity in a reactor core can be affected by the system pressure. The pressure coefficient of reactivity is defined as the change in reactivity per unit change in pressure. The pressure coefficient of reactivity for the reactor is the result of the effect of pressure on the density of the moderator. For this reason, it is sometimes referred to as the moderator density reactivity coefficient. As pressure increases, density correspondingly increases, which increases the moderator-to-fuel ratio in the core. In the typical under moderated core the increase in the moderator-to-fuel ratio will result in a positive reactivity addition. In reactors that use light water as a moderator, the absolute value of the pressure reactivity coefficient is seldom a major factor because it is very small compared to the moderator temperature coefficient of reactivity.

2.19.1.4 Void Coefficient

In systems with boiling conditions, such as boiling water reactors (BWR), the pressure coefficient becomes an important factor due to the larger density changes that occur when the vapor phase of water undergoes a pressure change. This is the void coefficient. The void coefficient is caused by the formation of steam voids in the moderator. The void coefficient of reactivity is defined as the change in reactivity per percent change in void volume. As the reactor power is raised to the point where the steam voids start to form, voids displace moderator from the coolant channels within the core. This displacement reduces the moderator-to-fuel ratio, and in an under moderated core, results in a negative reactivity addition, thereby limiting reactor power rise. The void coefficient is significant in water-moderated reactors that operate at or near saturated conditions.

2.19.1.5 Neutron Poison

A neutron poison is a substance with a large neutron absorption cross section and this affects the chain reaction. Some of the fission products generated during fission have a high neutron absorption capacity, such as xenon 135 (microscopic cross-section $\sigma = 2,000,000$ barns) and samarium 149 ($\sigma = 74,500$ b). Because these two fission product poisons remove neutrons from the reactor, they will have an impact on the thermal utilization factor and thus the reactivity. The poisoning of a reactor core by these fission products may become so serious that the chain reaction comes to a standstill.

Xenon-135 in particular has a tremendous impact on the operation of a nuclear reactor. The inability of a reactor to be restarted due to the buildup of xenon-135 (reaches a maximum after about 10 hours) is sometimes referred to as *xenon precluded start-up*. The period of time in which the reactor is unable to override the effects of xenon-135 is called the *xenon dead time* or *poison outage*. During periods of steady state operation, at a constant neutron flux level, the xenon-135 concentration builds up to its equilibrium value for that reactor power in about 40 to 50 hours. When the reactor power is increased, xenon-135 concentration initially decreases because the burn up is increased at the new higher power level. Thus, the dynamics of Xenon poisoning represents a positive reactivity feedback with importance for the stability of the flux pattern and geometrical power distribution especially in physically large reactors.

Because 95% of the xenon-135 production is from iodine 135 decay, which has a 6 to 7 hour half-life, the production of xenon-135 remains constant; at this point, the xenon-135 concentration reaches a minimum. The concentration then increases to the equilibrium for the new power level in the same time, roughly 40 to 50 hours. The magnitude and the rate of change of concentration during the initial 4 to 6 hour period following the power change is dependent upon the initial power level and on the amount of change in power level; the xenon-135 concentration change is greater for a larger change in power level. When reactor power is decreased, the process is reversed.

Because samarium-149 is not radioactive and is not removed by decay, it presents problems somewhat different from those encountered with xenon-135. The equilibrium concentration and (thus the poisoning effect) builds to an equilibrium value during reactor operation in about

500 hours (about three weeks), and since samarium-149 is stable, the concentration remains essentially constant during reactor operation. Another problematic isotope that is building up is gadalonium 157 with microscopic cross-section of $\sigma = 200{,}000$ b.

However neutron poisons, are intentionally inserted into some types of reactors in order to lower the high reactivity of their initial fresh fuel load. Fixed burnable poisons are generally used in the form of compounds of boron or gadalonium that are shaped into separate lattice pins or plates, or introduced as additives to the fuel. Some of these poisons deplete as they absorb neutrons during reactor operation, while others remain relatively constant. Soluble poisons, also called chemical shim, produce spatially uniform neutron absorption when dissolved in the water. The most common soluble poison in commercial PWR is boric acid, which is often referred to as soluble boron. The boric acid in the coolant decreases the thermal utilization factor, causing a decrease in reactivity. By varying the concentration of boric acid in the coolant, a process referred to as boration and dilution, the reactivity of the core can be easily varied. The variation in boron concentration allows control rod use to be minimized. Soluble poisons are also used in emergency shutdown systems. For accelerated shutdown the operators can inject solutions containing neutron poisons directly into the reactor coolant. Various solutions, including sodium polyborate and gadolinium nitrate ($Gd(NO_3)_3 \cdot xH_2O$), are used

2.20 REACTOR KINETICS

The response of neutron flux and reactor power to changes in reactivity is much different in a critical reactor than in a subcritical reactor. The reliance of the chain reaction on delayed neutrons makes the rate of change of reactor power controllable. The reactor period is defined as the time required for reactor power to change by a factor of "e," where "e" is the base of the natural logarithm and is equal to about 2.718. The reactor period is usually expressed in units of seconds. From the definition of reactor period, it is possible to develop the relationship between reactor power and reactor period that is expressed by

$$P = P_0 \cdot \exp(t/\tau)$$

where:

P = transient reactor power
P_0 = initial reactor power
τ = reactor period (seconds)
t = time during the reactor transient (seconds)

The smaller the value of τ, the more rapid the change in reactor power. If the reactor period is positive, reactor power is increasing. If the reactor period is negative, reactor power is decreasing. The first term in equation below is the prompt term and the second term is the delayed term.

$$\tau = \frac{l'}{\rho} + \frac{\overline{\beta}_{eff} - \rho}{\lambda_{eff}\rho + \dot{\rho}}$$

where:

l^* = prompt generation lifetime
$\overline{\beta}_{eff}$ = effective delayed neutration fraction
ρ = reactivity
λ_{eff} = effective delayed neutron precursor decay constant
$\dot{\rho}$ = rate of change of reactivity

Basic Physics of Nuclear Reactors

In order to understand the time-dependent behaviour of a reactor we need equations that describe the response of the prompt and delayed neutron populations to changes in reactivity. This problem is mathematically complex because the neutron population in a reactor is a function of both space (i.e., position in the core) and time. For many practical situations, we can assume that the spatial and temporal behaviour are separable. This allows us to write equations of reactor kinetics as a function of time alone. The space-independent equations of reactor kinetics, which are often called the "point kinetics" equations are:

$$dn(t)/dt = (\rho(t) - \beta)n(t)/l^* + \Sigma \lambda_i C_i(t)$$
$$dC_i(t)/dt = \beta_i n(t) - \lambda_i C_i(t)$$

where

$n(t)$ = is the reactor power,
$\rho(t)$ = is the net reactivity,
β = is the effective delayed neutron fraction,
l^* = is the prompt neutron life time,
λ_i = is the decay constant of the ith precursor group,
C_i = is the concentration of the ith precursor group, and
N = is the number of delayed neutron precursor groups

2.20.1 Effective Delayed Neutron Fraction

Recall that, the *delayed neutron fraction* is the fraction of all fission neutrons that are born as delayed neutrons. The value of depends upon the actual nuclear fuel used. The delayed neutron precursors for a given type of fuel are grouped on the basis of half-life. The following table lists the fractional neutron yields for each delayed neutron group of three common types of fuel.

Table 2.5: Delayed Neutron Fractions for Fuels

Group	Half-Life (sec)	Uranium-235	Uranium-238	Plutonium-239
1	55.6	0.00021	0.0002	0.00021
2	22.7	0.00141	0.0022	0.00182
3	6.22	0.00127	0.0025	0.00129
4	2.30	0.00255	0.0061	0.00199
5	0.61	0.00074	0.0035	0.00052
6	0.23	0.00027	0.0012	0.00027
Total	-	0.00650	0.0157	0.00200

The term β_{avg} (pronounced beta) is the *average delayed neutron fraction*. The value of β_{avg} is the weighted average of the total delayed neutron fractions of the individual types of fuel. Each total delayed neutron fraction value for each type of fuel is weighted by the percent of total neutrons that the fuel contributes through fission. If the percentage of fissions occurring in the different types of fuel in a reactor changes over the life of the core, the average delayed neutron fraction will also change. For a light water reactor using low enriched fuel, the average delayed neutron fraction can change from 0.0070 to 0.0055 as uranium-235 is burned out and plutonium-239 is produced from uranium-238.

Delayed neutrons do not have the same properties as prompt neutrons released directly from fission. The average energy of prompt neutrons is about 2 MeV. This is much greater than the average energy of delayed neutrons (about 0.5 MeV). The fact that delayed neutrons are born at lower energies has two effects. First, delayed neutrons have a much lower probability of causing fast fissions than prompt neutrons because their average energy is less than the minimum required for fast fission to occur. Second, delayed neutrons have a lower probability of leaking out of the core while they are at fast energies, because they are born at lower energies and subsequently travel a shorter distance as fast neutrons. These two considerations (lower fast fission factor and higher fast non-leakage probability for delayed neutrons) are taken into account by a term called the *importance factor* (I). The importance factor relates the average delayed neutron fraction to the effective delayed neutron fraction.

The *effective delayed neutron fraction* is defined as the fraction of neutrons at thermal energies which were born delayed. The effective delayed neutron fraction is the product of the average delayed neutron fraction and the importance factor.

$$\beta_{eff} = \beta_{avg} \cdot I$$

In a small reactor with highly enriched fuel, the increase in fast non-leakage probability will dominate the decrease in the fast fission factor, and the importance factor will be greater than one. In a large reactor with low enriched fuel, the decrease in the fast fission factor will dominate the increase in the fast non-leakage probability and the importance factor will be less than one (about 0.97 for a commercial PWR).

2.21 REACTOR CONTROL

In a nuclear reactor, fuel is inside the reactor, unlike the fossil fuelled plants where it flows into the boiler from outside. Hence it is difficult in most cases to continuously add fuel. Hence additional fuel (excess reactivity) is kept so that continued operation is possible for the designed duration without need to refuel. Also excess reactivity is needed for overriding the effect of neutron poisons. Hence there is an excess reactivity built in the reactor. Its value will come down as the fuel gets used up. To override this excess reactivity, all reactors are provided with control rods made of neutron absorbing materials. The control rods are also needed to stop the fission chain reaction when needed to shutdown the reactor. Since the power level of the reactor is proportional to the neutron flux, the reactor control is achieved by making the neutron multiplication factor well below 1. This is achieved by lowering the control rods gradually. In case of an emergency the control rods are dropped into the core.

Reactor power control is also possible through addition or removal of moderator. In Pressurised Heavy water (CANDU) reactors, moderator level control is used in older designs, while in later designs neutron absorber material in the form of Boric acid is added to the moderator or removed from the moderator stream.

Neutron absorber materials generally used are boron and cadmium. Boron has two isotopes B10 and B11. Natural boron contains 20% of B10, which is the neutron absorber. It is used in the form of carbide of boron. When any neutron is absorbed in boron a (n, α) reaction occurs. This means generation of He which goes to swell the material.

2.22 BIOLOGICAL EFFECTS OF RADIATION & SHIELDING

Whether the source of radiation is natural or man-made, whether it is a small dose of radiation or a large dose, there will be some biological effects. Although we tend to think of biological effects in terms of the effect of radiation on living cells, in actuality, ionizing radiation, by definition, interacts only with atoms by a process called ionization. Thus, all biological damage effects begin with the consequence of radiation interactions with the atoms forming

Basic Physics of Nuclear Reactors

the cells. However, each cell as is the case for the human body, is mostly water. When radiation interacts with water, it may break the bonds that hold the water molecule together, producing fragments such as hydrogen (H) and hydroxyls (OH). These fragments may recombine or may interact with other fragments or ions to form compounds, such as water, which would not harm the cell. However, they could combine to form toxic substances, such as hydrogen peroxide (H_2O_2), which can contribute to the destruction of the cell. Not all living cells are equally sensitive to radiation. Those cells which are actively reproducing are more sensitive than those which are not. As a result, living cells can be classified according to their rate of reproduction, which also indicates their relative sensitivity to radiation. This means that different cell systems have different sensitivities. Lymphocytes (white blood cells) and cells which produce blood are constantly regenerating, and are, therefore, the most sensitive. Reproductive and gastrointestinal cells are not regenerating as quickly and are less sensitive. The nerve and muscle cells are the slowest to regenerate and are the least sensitive cells.

In view of the biological effects, it is essential to provide shielding materials that would stop the radiation from affecting the occupational workers in nuclear facilities. Different shielding materials are used depending on whether it is neutron, alpha, beta or gamma.

Shielding materials commonly used are stainless steel, lead, tantalum, tungsten, water, Borated concrete and Boron compounds. The heavy elements attenuate the gamma radiation and slow down fast neutrons, while hydrogenous materials moderate the neutrons and boron compounds capture the neutrons.

SUMMARY

Basic physics of atoms and their constituents is essential to the understanding of the fission process. This chapter has attempted to give a overview of the different aspects of neutron interactions and explained some of the scientific terms used in the nuclear industry. This is essential for better understanding of the remaining chapters.

BIBLIOGRAPHY

1. Reactor Theory, Department of Energy Fundamentals Handbook. DOE-HDBK-1019/2-93(1993).
2. S.Glasstone and A.Sesonske, Nuclear Reactor Engineering, #rd Edition, Von Nostrand (1981).
3. John R. Lamarsh, Introduction to Nuclear Engineering, Addison Wesley Publishing company (1983).
4. S.Garg, F.Ahmed and L.S.Kothari, Physics of Nuclear Reactors, Tata McGraw Hill, New Delhi (1986).
5. S.N. Ghoshal, Nuclear Physics, S.Chand Publishers, Delhi (2005).
6. Kenneth D. Kok (ed.), Handbook of Nuclear Engineering, CRC Press, (2009).
7. Dan Gabriel Cacuci (ed.), Handbook of Nuclear Engineering, Springer (2010).
8. Steven B., Krivit, Jay H. Lehr, Thomas B. Kingery, Nuclear Energy Encyclopedia: Science, Technology, and Applications, John Wiley (2011).

ASSIGNMENTS

1. A nuclear reactor has a heat generation of 200 MW. How many atoms of U235 will be needed to fissioned per second. If the fuel core contained 60 Kg of U235, what fraction should have been used up after 50 days of operation.
2. What is Binding Energy? What is its importance with reference to fission and fusion?
3. Determine the number eV per second consumed by a 100 W lamp.

4. What is critical mass?
5. What do you mean by the term Cross section? What are its units?
6. Explain the neutron life cycle.
7. What are prompt and delayed neutrons? What is the role played by delayed neutrons in the fission process?
8. What is reactivity? Briefly explain the different reactivity coefficients encountered in nuclear reactors.
9. What is a moderator? Compare the different moderator materials.
10. What is decay heat in a nuclear reactor? Why do we not talk about decay heat with reference to fossil fuelled plants.
11. Explain Prompt criticality. What are its consequences?
12. Differentiate between a breeder, transmuter and converter.

3

Basics of Power Plant

3.0 INTRODUCTION

Thermodynamics is the science of many processes involved in one form of energy being changed into another. It is a set of principles that enable us to understand and follow energy as it is transformed from one form to other. The zeroth law of thermodynamics was enunciated after the first law. It states that if two bodies are each in thermal equilibrium with a third, they must also be in thermal equilibrium with each other. Equilibrium implies the existence of a situation in which the system undergoes no net change, and there is no net transfer of heat between the bodies.

The first law of thermodynamics says that energy can't be destroyed or created. When one energy form is converted into another, the total amount of energy remains constant. An example of this law is a gasoline engine. The chemical energy in the fuel is converted into various forms including kinetic energy of motion, potential energy, chemical energy in the carbon dioxide, and water of the exhaust gas.

The second law of thermodynamics is the entropy law, which says that all physical processes proceed in such a way that the availability of the energy involved decreases. This means that no transformation of energy resource can ever be 100% efficient. Entropy is a measure of disorder or chaos, when entropy increases disorder increases.

The third law of thermodynamics is the law of unattainability of absolute zero temperature, which says that entropy of an ideal crystal at zero degrees Kelvin is zero. It's unattainable because it is the lowest temperature that can possibly exist and can only be approached but not actually reached. This law is not needed for most thermodynamic work, but is a reminder that like the efficiency of an ideal engine, there are absolute limits in physics.

The steam power plants works on modified rankine cycle in the case of steam engines and isentropic cycle (Constant Entropy) concerned in the case of impulse and reaction steam turbines. In the case of I.C. Engines (Diesel Power Plant) it works on Otto cycle, diesel cycle or dual cycle and in the case of gas turbine it works on Brayton cycle. However in the case of non-conventional energy generation it is complicated and depends upon the type of the system viz., thermo electric or thermionic basic principles and theories. Steam plant employed in nuclear plants utilize the modified rankine cycle, while gas cooled reactors use the Brayton cycle. This chapter discusses the rankine and Brayton cycles used in nuclear power plants.

3.1 CARNOT CYCLE

This cycle is of great value to heat power theory although it has not been possible to construct a practical plant on this cycle. It has high thermodynamics efficiency. A schematic of carnot cycle with air as the fluid is given in figure 3.1.

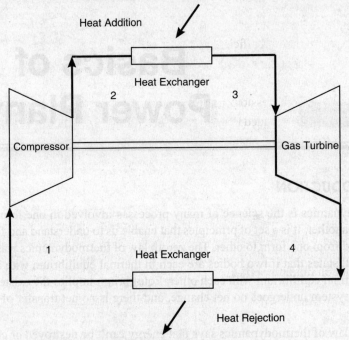

Fig. 3.1: Carnot Cycle Schematic

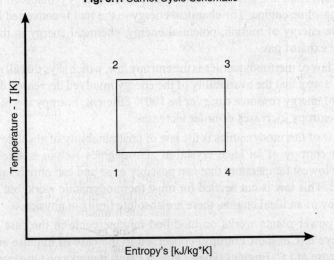

Fig. 3.2: Carnot Cycle on T-S Diagram

Here air enters at point 1 and exits the compressor at point 2. The compression is considered as isentropic or constant entropy process. Then air enters a heat exchanger and exits at point 3. The heat exchange is considered to take place at constant temperature. Then air enters a turbine and exits at point 4. Here the process is again considered isoentropic. Then air enters a heat exchanger where it rejects heat to a sink at constant temperature and returns to point 1.

Basics of Power Plant

Fig 3.2 shows the carnot cycle on a Temperature Entropy diagram. The thermal efficiency can be given as a function of the specific external work (specific net power output) and heat added to the working fluid as given below:

$$\eta = w/q_h = (w_t - w_c)/q_h = (q_h - q_r)/q_h = 1 - q_r/q_h = 1 - T_1 \Delta s / T_2 \Delta s$$
$$= 1 - T_1/T_2$$

Where

η = thermal efficiency
w = specific external work (specific net power output)
w_t = expansion specific power output
w_c = compression specific power output
q_h, q_r = heat added to and heat rejected from the working fluid
Δs = entropy change during heat addition and rejection
T_1, T_2 = Compressor inlet outlet temperatures
T_3, T_4 = Turbine inlet outlet temperatures

For isentropic expansion

$$T_2/T_1 = (p_2/p_1)^{(\gamma-1)/\gamma} = T_3/T_4 = (p_3/p_4)^{(\gamma-1)/\gamma}$$

As γ for air or a perfect gas like steam is 1.4, the ratio of temperatures is a constant value. Hence η of carnot cycle is dependant only on the temperatures at which heat is added and rejected and not on fluid properties.

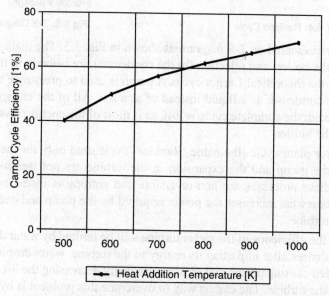

Compressor Inlet Temperature: 298 [K]

Fig. 3.3: Carnot Cycle efficiency vs Heat addition temperature

Fig. 3.3 shows the effect of increase in the heat addition temperature for a given heat rejection temperature. The heat rejection temperature is taken as 298 K being the possible low temperature one can reach. The behavior of steam which is a gas is akin to the behavior of air. The highest temperature reached is limited by material strength and corrosion considerations.

3.2 RANKINE CYCLE

Steam engine and steam turbines in which steam is used as working medium follow Rankine cycle. The Rankine cycle is sometimes referred to as a practical carnot cycle. This cycle can be carried out in four pieces of equipment connected by pipes for conveying working medium as shown in Fig. 3.4. Here also the water is compressed in the pump and fed to the steam generator where it receives heat from the boiler and produces steam. The boiler can receive the heat from either nuclear fission or by burning coal, oil or natural gas. This steam is fed to the Turbine where the steam power rotates the turbine which in turn rotates a generator to produce electricity. The fluid exiting the turbine is brought to the lowest temperature in a condenser.

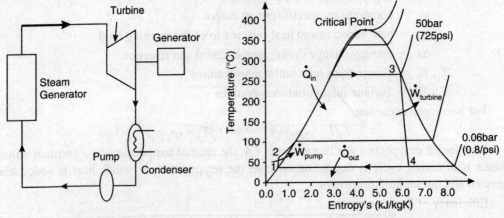

Fig. 3.4: Rankine Cycle   Fig 3.5: T-S Diagram

The cycle is represented on T-S diagram as shown in Fig. 3.5. The main difference is that heat addition (in the boiler) and rejection (in the condenser) are isobaric in the Rankine cycle and isothermal in the theoretical Carnot cycle. A pump is used to pressurize the working fluid received from the condenser as a liquid instead of as a gas. All of the energy in pumping the working fluid through the complete cycle is lost, as is most of the energy of vaporization of the working fluid in the boiler.

In a real power plant cycle (the name 'Rankine' cycle used only for the ideal cycle), the compression by the pump and the expansion in the turbine are not isentropic (Figure 3.6). In other words, these processes are non-reversible and entropy is increased during the two processes. This somewhat increases the power required by the pump and decreases the power generated by the turbine.

In particular the efficiency of the steam turbine will be limited by water droplet formation. As the steam condenses after imparting its energy to the turbine, water droplets hit the turbine blades at high speed causing pitting and erosion, gradually decreasing the life of turbine blades and efficiency of the turbine. The easiest way to overcome this problem is by superheating the steam. On the T-S diagram above, state 3 is above a two phase region of steam and water so after expansion the steam will be very wet. By superheating, state 3 will move to the right of the diagram and hence produce a drier steam after expansion. All fossil fuelled plants produce steam at superheated conditions (160 b, 540° C). Pressurised water reactors, Boiling water reactors and Heavy water reactors work on the saturated steam cycle, while Fast Reactors cooled by sodium work on the Superheated cycle as in a fossil fuel power plant.

Basics of Power Plant

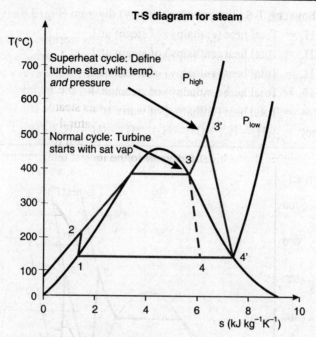

Fig. 3.6: Practical Rankine cycle

Efficiency of Rankine cycle = $(H_1 - H_2)/(H_1 - Hw_2)$
where,

H_1 = Total heat (enthalpy) of steam at entry pressure
H_2 = Total heat of steam at condenser pressure (exhaust pressure)
Hw_2 = Total heat of water at exhaust pressure

3.3 REHEAT CYCLE

In this cycle steam is extracted from a suitable point in the turbine and reheated generally to the original temperature by flue gases. Reheating is generally used when the pressure is high say above 100 kg/cm². The various advantages of reheating are as follows:

(i) It increases dryness fraction of steam at exhaust so that blade erosion due to impact of water particles is reduced.
(ii) It increases thermal efficiency.
(iii) It increases the work done per kg of steam and these results in reduced size of boiler.

The disadvantages of reheating are as follows:

(i) Cost of plant is increased due to the reheater and its long connections.
(ii) It increases condenser capacity due to increased dryness fraction.

Fig. 3.7 shows flow diagram of reheat cycle. First turbine is high-pressure turbine and second turbine is low pressure (L.P.) turbine.

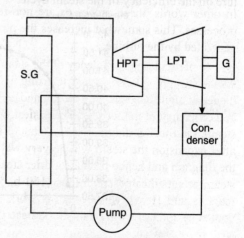

Fig 3.7: Reheat Cycle

This cycle is shown on T-S (Temperature entropy) diagram (Fig. 3.8).

If, H_1 = Total heat (enthalpy) of steam at 1
H_2 = Total heat (enthalpy) of steam at 2
H_3 = Total heat (enthalpy) of steam at 3
H_4 = Total heat (enthalpy) of steam at 4
Hw_4 = Total heat (enthalpy) of water at 4

Efficiency = $\{(H_1 - H_2) + (H_3 - H_4)\}/\{H_1 + (H_3 - H_2) - Hw_4\}$

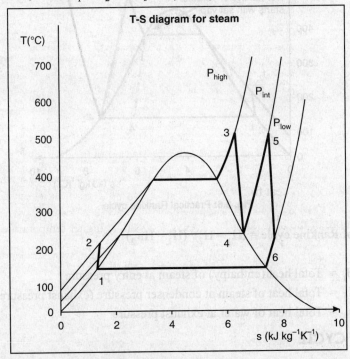

Fig 3.8: T-S diagram of Reheat Cycle

Figure 3.9 shows the impact of reheat pressure and correspondingly higher reheat temperature on the efficiency of the steam cycle.

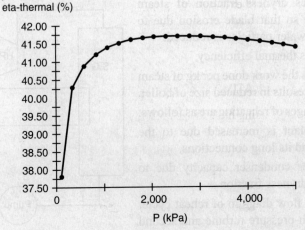

Fig. 3.9: Rankine Cycle with reheat Improving Efficiency with Reheat Pressure

Basics of Power Plant

3.4 REGENERATIVE CYCLE (FEED WATER HEATING)

The process of extracting steam from the turbine at certain points during its expansion and using this steam for heating for feed water is known as Regeneration or Bleeding of steam. The arrangement of bleeding the steam at two stages is shown in Fig. 3.10.

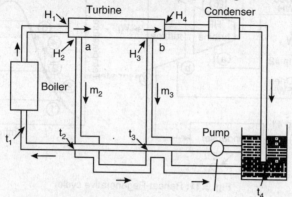

Fig. 3.10: Regenerative feed cycle

The regenerative features here effectively raise the nominal cycle heat input temperature, by reducing the addition of heat from the boiler/fuel source at the relatively low feedwater temperatures that would exist without regenerative feedwater heating. This improves the efficiency of the cycle, as more of the heat flow into the cycle occurs at higher temperature

Let,

m_2 = Weight of bled steam at a per kg of feed waterheated

m_3 = Weight of bled steam at b per kg of feed water heated

H_1, Hw_1 = Enthalpies of steam and water in boiler

H_2, H_3 = Enthalpies of steam at points a and b

t_2, t_3 = Temperatures of steam at points a and b

H_4, Hw_4 = Enthalpy of steam and water exhausted to hot well.

Work done in turbine per kg of feed water between entrance and $a = H_1 - H_2$

Work done between a and $b = (1 - m_2)(H_2 - H_3)$

Work done between b and exhaust $= (1 - m_2 - m_3)(H_3 - H_4)$

Total heat supplied per kg of feed water $= H_1 - Hw_2$

Efficiency (η) = Total work done/Total heat supplied

$= \{(H_1 - H_2) + (1 - m_2)(H_2 - H_3) + (1 - m_2 - m_3)(H_3 - H_4)\}/(H_1 - Hw_2)$

3.5 REHEAT-REGENERATIVE CYCLE

In steam power plants using high steam pressure combined reheat and regenerative cycle is used. The thermal efficiency of this cycle is higher than the plain reheat or regenerative cycle. Fig. 3.11 shows the flow diagram of reheat regenerative cycle. This cycle is commonly used to produce high pressure steam (> 90 kg/cm^2) to increase the cycle efficiency.

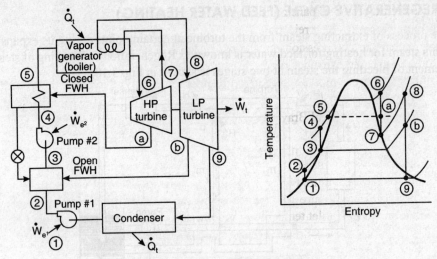

Fig. 3.11: Reheat-Regenerative cycle

3.6 BRAYTON CYCLE

The term Brayton cycle has more recently been given to the gas turbine engine (Fig. 3.12). This also has three components:
- a gas compressor
- a burner (or combustion chamber)
- an expansion turbine

Ideal Brayton cycle includes

Isentropic process: Ambient air is drawn into the compressor, where it is pressurized.

Isobaric process: The compressed air then runs through a combustion chamber, where fuel is burned, heating that air—a constant-pressure process, since the chamber is open to flow in and out.

Isentropic process: The heated, pressurized air then gives up its energy, expanding through a turbine (or series of turbines). Some of the work extracted by the turbine is used to drive the compressor.

Isobaric process: Heat rejection (in the atmosphere).

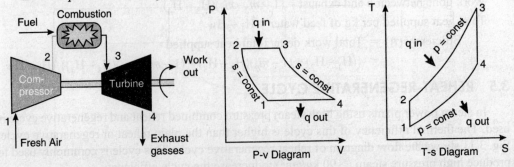

Fig. 3.12: Brayton cycle

Actual Brayton cycle has:
- adiabatic process — compression.
- isobaric process — heat addition.

—adiabatic process —expansion.
—isobaric process —heat rejection.

Since neither the compression nor the expansion can be truly isentropic, losses through the compressor and the expander represent sources of inescapable working inefficiencies. In general, increasing the compression ratio is the most direct way to increase the overall power output of a Brayton system.

The efficiency of the ideal Brayton cycle is

$$\eta = 1 - \frac{T_1}{T_2} = 1 - \left(\frac{P_1}{P_2}\right)^{(\gamma-1)/\gamma}$$

where γ is the specific heat ratio. Figure 3.13 indicates how the cycle efficiency changes with an increase in pressure ratio. Figure 3.14 indicates how the specific power output changes with an increase in the gas turbine inlet temperature for two different pressure ratio values.

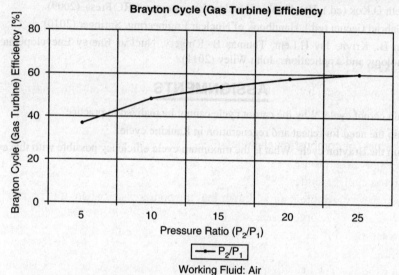

Working Fluid: Air
Compressor Inlet Temperature: 298 [K]— Gas Turbine Inlet Temperature: 1,500 [K]
Fig. 3.13: Brayton Cycle Efficiency vs Pressure Ratio

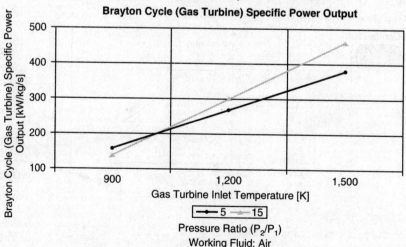

Pressure Ratio (P_2/P_1)
Working Fluid: Air
Compressor Inlet Temperature: 298 [K]
Fig. 3.14: Brayton cycle efficiency vs Gas Inlet Temperature

SUMMARY

This chapter has introduced to the carnot, rankine cycle along with the concept of reheat and regeneration to improve cycle efficiencies. Higher the efficiency, less heat is required to produce the same electrical power. This in turn results in fewer amounts of fuel and less amount of waste.

BIBLIOGRAPHY

1. P.K.Nag, Power Plant Engineering, Tata McGraw Hill Second Edition. (2001).
2. P.K.Nag, Engineering Thermodynamics, Tata McGraw Hill, (2003).
3. Domkundwar, Power plant engineering, Dhanpat Rai Publication (2002).
4. Cengel, Thermodynamics (SI Units) 5th Edition, Tata McGraw Hill, (2003).
5. P.L. Ballaney, Thermal Engineering, Khanna Publishers, Delhi, (1995)
6. Kenneth D.Kok (ed.), Handbook of Nuclear Engineering, CRC Press, (2009).
7. Dan Gabriel Cacuci (ed.), Handbook of Nuclear Engineering, Springer (2010).
8. Steven B., Krivit, Jay H.Lehr, Thomas B. Kingery, Nuclear Energy Encyclopedia: Science, Technology, and Applications, John Wiley (2011).

ASSIGNMENTS

1. Explain carnot cycle. Why the carnot cycle cannot be realized in practice?
2. Explain the need for reheat and regeneration in Rankine cycle.
3. Explain the Brayton cycle. What is the maximum cycle efficiency possible with this cycle?

4

Reactor Fuel Cycle

4.0 INTRODUCTION

The production of the reactor fuel, its utilization in the reactor and recovery of the unused fissile and fertile materials from the spent fuel constitute the fuel cycle. Uranium the only naturally occurring nuclear fissile material is a slightly radioactive metal that occurs throughout the Earth's crust. It is about 500 times more abundant than gold and about as common as tin. It is present in most rocks and soils as well as in many rivers and sea water. It is, for example, found in concentrations of about four parts per million (ppm) in granite, which makes up 60% of the Earth's crust. In fertilizers, uranium concentration can be as high as 400 ppm (0.04%), and some coal deposits contain uranium at concentrations greater than 100 ppm (0.01%). There are a number of areas around the world where the concentration of uranium in the ground is sufficiently high that extraction of it for use as nuclear fuel is economically feasible. Such concentrations are called ore.

Fuel needed by different reactors include uranium oxide, uranium carbide, and uranium metal and in some cases enriched U235. The enrichments vary from 2.5 to 85%. In view of this process to separate the isotopes of U235 and U238 have been developed. After its utilization in the reactor, the fuel would contain unused Uranium, Plutonium produced by conversion of U238, fission products etc. The recovery of unused fuel materials is a complex chemical process referred to as reprocessing. In view of the presence of slowly decaying fission products, a certain time is required for the spent fuel from the reactor to come down both in radioactivity and temperature before it can be reprocessed. Research is continuing to develop reprocessing methods for short cooled fuels. In brief, fuel cycle includes the fuel supply, fabrication, generation, fuel storage, reprocessing, recycling, waste management, disposal and decommissioning. It will be necessary to simplify the fuel cycle to reduce costs, while still minimizing the environmental impact, Research goal is to improve fuel performance for longer life in the reactor and the development of advanced fuels.

The method of separation of fission products assumes importance especially for thermal reactors, where these may act as neutron poisons. If the fuel is to be used in a Fast reactor, then this is not a problem as absorption cross sections for neutrons by these poisons is less for high energy neutrons. Last but not the least is the problem of disposal of the waste containing the radioactive fission products. A schematic of the nuclear fuel cycle is depicted in Fig. 4.1. The part of the fuel cycle from Mining to power plant is called the front end of the fuel cycle, while the part from power plant to waste management is called the back end of the fuel cycle. A more detailed route of the fuel is depicted in figure 4.2 for better understanding. This chapter is devoted to an overview of the reactor fuel cycle.

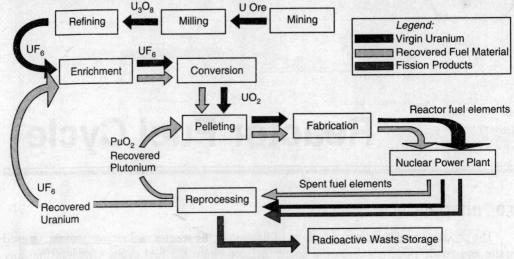

Fig. 4.1: Schematic of Nuclear Fuel Cycle

It is essential to understand the full cycle as closure of the cycle is essential for sustainability of nuclear power. Many countries, especially USA have not gone in for reprocessing from considerations of proliferation. However reprocessing is must to extract the unutilized fuel and breed fuel to effectively utilize our natural uranium resources.

4.1 MATERIAL BALANCE IN THE NUCLEAR FUEL CYCLE

The following gives an idea of the quantities in the nuclear fuel cycle. These are not accurate but may be regarded as typical for the operation of a 1000 MWe nuclear power reactor:

Mining 20 000 tones of 1% uranium ore.

Milling - 230 tonnes of uranium oxide concentrate (with 195 t U)

Conversion - 288 tonnes UF_6 (with 195 t U)

Enrichment - 35 tonnes UF_6 (with 24 t enriched U) - balance is 'tails'

Fuel fabrication - 27 tonnes UO_2 (with 24 t enriched U)

Reactor operation - 8640 million kWh (8.64 TWh) of electricity at full output.

Used fuel - 27 tonnes containing 240kg plutonium, 23 t uranium (0.8% U-235), 720kg fission products.

4.2 URANIUM MINING

In general, open pit mining is used where deposits are close to the surface and underground mining is used for deep deposits, typically greater than 120 meters deep. Open pit mines require large holes on the surface, larger than the size of the ore deposit, since the walls of the pit must be sloped to prevent collapse. As a result, the quantity of material that must be removed in order to acess the ore (Fig. 4.3) may be large. Underground mines have relatively small surface disturbance and the quantity of material that must be removed to access the ore is considerably less than in the case of an open pit mine. An increasing proportion of the world's uranium now comes from in situ leaching (ISL), where oxygenated groundwater is circulated through a very porous ore body to dissolve the uranium and bring it to the surface. ISL may be with

Reactor Fuel Cycle

slightly acid or with alkaline solutions to keep the uranium in solution. The uranium is then recovered from the solution as in a conventional mill. In the case of undergroundd uranium mines, special precautions, consisting primarily of increased ventilation, are required to protect against airborne radiation exposure.

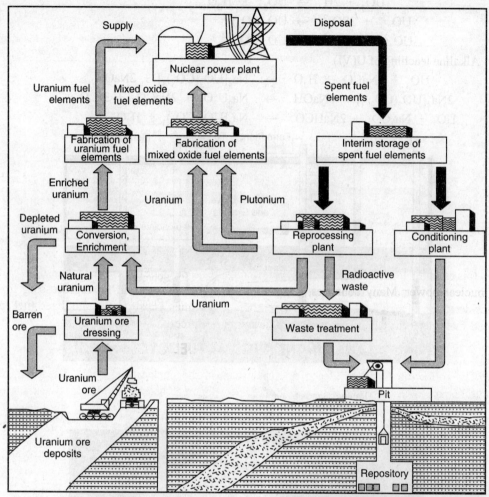

Fig. 4.2: Uranium Mining to Waste Repository

4.3 URANIUM MILLING

A uranium mill is a chemical plant that extracts uranium from mined ore. At conventional mills, the ore arrives via truck and is crushed, ground and leached. In most cases, sulfuric acid is the leaching agent, but alkaline leaching can also be done. The leaching agent extracts some 90 to 95 percent of the uranium from the ore. This is done using either acid or alkali and an oxidant to convert U (IV) to U (VI).

Fig. 4.3: Uranium Ore

Oxidation of U(IV) to U(VI)

$$UO_2 + 2Fe^{3+} \rightarrow UO_2^{2+} + 2Fe^{2+}$$

Acid leaching of U(VI)

$$UO_3 + 2H^+ \rightarrow UO_2^{2+} + H_2O$$
$$UO_2^{2+} + 3SO_3^{2-} \rightarrow UO_2(SO_4)_3^{4-}$$
$$UO_2^{2+} + 3CO_3^{2-} \rightarrow UO_2(CO_3)_3^{4-}$$

Alkaline leaching of U(VI)

$$UO_3 + 3NaCO_3 + H_2O \rightarrow Na_4[UO_2(CO_3)_3] + 2NaOH$$
$$2Na_4[UO_2(CO_3)_3] + 6NaOH \rightarrow Na_2U_2O_7 + 2NaCO_3 + 3H_2O$$
$$UO_3 + Na_2CO_3 + 2NaHCO_3 \rightarrow Na_4[UO_2(CO_3)_3] + H_2O$$

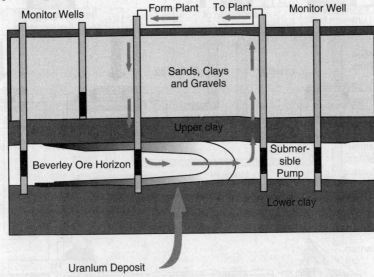

Fig. 4.4: Uranium Mining

Fig. 4.5: Yellow Cakes

The uranium is then removed from this solution, precipitated as uranium oxide concentrate (U_3O_8) and dried. It is sometimes referred to as yellowcake (Fig 4.5) and generally contains more than 80% uranium. The original ore may contain as little as 0.1% uranium or less. The leach solution is pumped from the ore body underground to the processing plant, and ion exchange separates the uranium from the solution. About 200 t of yellow cake are required for a 1000 MWe nuclear power plant to generate electricity for 1 year.

4.4 CONVERSION TO URANIUM HEXAFLUORIDE

The product of a uranium mill is not directly usable as a fuel for a nuclear power reactor. Additional processing, generally referred to as enrichment is required for most kinds of reactors. This process requires uranium to be in gaseous form and the way this is achieved is to convert it to uranium hexafluoride (UF_6), which is a gas at relatively low temperatures. The UF_6 is then pressurized and cooled to a liquid. In its liquid state it is drained into cylindrical tanks where it solidifies after cooling. The UF_6 in the solid form is then shipped to an enrichment plant.

4.5 ENRICHMENT

Natural uranium consists, primarily, of a mixture of two isotopes (atomic forms) of uranium. The fissile isotope of uranium is uranium-235 (235U); the remainder is mostly uranium-238 (238U). The former is only 0.7% while the latter constitutes remaining 99.3%. In Light Water nuclear reactors, a higher-than-natural concentration of 235U is required. The enrichment process produces this higher concentration, typically between 3.5% and 5% 235U, by removing 238U. This is done by separating gaseous uranium hexafluoride into two streams, one being enriched to the required level and known as low-enriched uranium. The other stream is progressively depleted in 235U and is called 'tails'.

There are two enrichment processes in large-scale commercial use, each of which uses uranium hexafluoride as feed: gaseous diffusion and gas centrifuge. They both use the physical properties of molecules, specifically the 1% mass difference, to separate the isotopes. The product of this stage of the nuclear fuel cycle is enriched uranium hexafluoride, which is reconverted to produce enriched uranium oxide. Other methods based on curved nozzle separation and laser enrichment have also been explored. The capital costs of enrichment plants are relatively high; around 6% of the total generation cost.

4.5.1 Gaseous diffusion

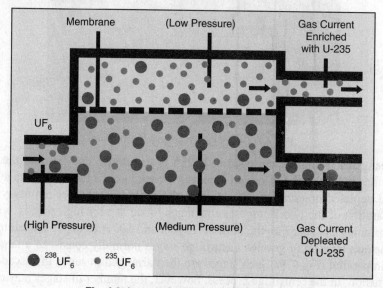

Fig. 4.6: Isotopic Separation by Gas Diffusion

In the gaseous diffusion enrichment plant, the solid uranium hexafluoride (UF_6) from the conversion process is heated until it becomes a liquid. The container becomes pressurized as the solid melts and UF_6 gas fills the top of the container. The UF_6 gas is slowly fed into the plant's pipelines where it is pumped through special filters called barriers or porous membranes (Figure 4.6). The holes in the barriers are so small that there is barely enough room for the UF_6 gas molecules to pass through. The isotope enrichment occurs when the lighter UF_6 gas molecules (with the U-234 and U-235 atoms) tend to diffuse faster through the barriers than the heavier UF_6 gas molecules containing U-238. One barrier isn't enough and it takes many hundreds of barriers, one after the other, before the UF_6 gas contains enough uranium-235 to be used in reactors. At the end of the process, the enriched UF_6 gas is withdrawn from the pipelines and condensed back into a liquid that is poured into containers. The UF_6 is then allowed to cool and solidify before it is transported to fuel fabrication facilities where it is turned into fuel assemblies for nuclear power reactors.

4.5.2 Gas centrifuge

The gas centrifuge uranium enrichment process uses a large number of rotating cylinders in series and parallel formations. Centrifuge machines are interconnected to form trains and cascades. In this process, UF_6 gas is fed into a series of vacuum tubes, each containing a rotor about one to two meters long and 15-20 cm diameter. When the rotors are spun rapidly at 50,000 to 70,000 rpm, the heavier molecules with U-238 increase in concentration towards the cylinder's outer edge. There is a corresponding increase in concentration of U-235 molecules near the centre. These concentration changes are enhanced by inducing the gas to circulate axially within the cylinder.

Fig. 4.7: Centrifuges in an enrichment plant

The stream that is slightly enriched in U-235 is withdrawn and fed into the next higher stage, while the slightly depleted stream is recycled back into the next lower stage. Eventually enriched and depleted uranium are drawn from the cascade at the desired assays. To obtain efficient separation of the two isotopes, centrifuges rotate at very high speeds, with the outer wall of the spinning cylinder moving at between 400 and 500 meters per second to give a million times the acceleration of gravity. Centrifuge stages normally consist of a large number of centrifuges in parallel (Fig 4.7). Such stages are then arranged in series in cascades similar to those for diffusion. In the centrifuge process, however, the number of stages may only be 10 to 20, instead of a thousand or more for diffusion.

4.6 FUEL FABRICATION

Nuclear fuel fabrication makes the enriched UF_6 into fuel for nuclear power reactors. Reactor fuel is generally in the form of ceramic pellets. These are formed from pressed uranium oxide that is sintered (baked) at a high temperature (over 1400° C). The pellets are then encased in metal tubes to form fuel rods (Figure 4.8), which are then arranged into a fuel assembly and ready for introduction into a reactor. The dimensions of the fuel pellets and other components of the fuel assembly are precisely controlled to ensure consistency in the characteristics of fuel bundles. In a fuel fabrication plant, great care is taken with the size and shape of processing vessels to avoid criticality (a limited chain reaction releasing radiation).

Fig. 4.8: Fuel Pins Assembly

4.7 FUEL IN POWER GENERATION

Inside a nuclear reactor the nuclei of uranium-235 atoms split (fission) and, in the process, release energy. This energy is used to heat water and turn it into steam. The steam is used to drive a turbine connected to a generator that produces electricity. Some of the uranium-238 in the fuel is turned into plutonium in the reactor core. The main plutonium isotope is also fissile and it yields about one third of the energy in a typical nuclear reactor.

An issue in operating reactors and hence specifying the fuel for them is fuel burn-up. This is measured in gigawatt-days per tonne and its potential is proportional to the level of enrichment. Hitherto a limiting factor has been the strength of the structural materials of fuel assemblies under high temperatures and irradiation conditions, and hence burn-up levels of about 40 GWd/t have required only around 4% enrichment. But with better materials and fuel, 55 GWd/t is possible (with 5% enrichment), and 70 GWd/t is in sight, though this would require 6% enrichment. The benefit of this is that operation cycles can be longer – around 24 months – and the number of fuel assemblies discharged as used fuel can be reduced by one third.

4.8 TRANSPORT OF RADIOACTIVE MATERIALS

Transport is an integral part of the nuclear fuel cycle. There are nuclear power reactors in operation in several countries but uranium mining is viable in only a few areas. Also, in the course of over forty years of operation by the nuclear industry, a number of specialized facilities have been developed in various locations around the world to provide fuel cycle services and there is a need to transport nuclear materials to and from these facilities. Most transports of nuclear fuel material occur between different stages of the cycle, but occasionally a material may be transported between similar facilities. With some exceptions, nuclear fuel cycle materials are transported in solid form, the exception being uranium hexafluoride (UF_6) which is considered a gas. Most of the material used in nuclear fuel is transported several times during the cycle.

Since nuclear materials are radioactive, it is important to ensure that radiation exposure of both those involved in the transport of such materials and the general public along transport routes is limited. Packaging for nuclear materials includes, where appropriate, shielding to reduce potential radiation exposures. In the case of some materials, such as fresh uranium fuel assemblies, the radiation levels are negligible and no shielding is required. Other materials, such as spent fuel and high-level waste, are highly radioactive and require special handling. To limit the risk in transporting highly radioactive materials, containers known as spent nuclear fuel shipping casks are used which are designed to maintain integrity under normal transportation conditions and during hypothetical accident conditions.

4.9 SPENT FUEL STORAGE

Fig. 4.9: Interim Fuel Storage

After its operating cycle, the reactor is shut down for refueling. The fuel discharged at that time (spent fuel) is stored either at the reactor site (commonly in a spent fuel pool) or potentially in a common facility away from reactor sites. If on-site pool storage capacity is exceeded, it may be desirable to store the now cooled aged fuel in modular dry storage facilities known as Independent Spent Fuel Storage Installations (ISFSI) at the reactor site or at a facility away from the site. The spent fuel rods are usually stored in water or boric acid (Figure 4.9), which provides both cooling (the spent fuel continues to generate decay heat as a result of residual radioactive decay) and shielding to protect the environment from residual ionizing radiation, although after at least a year of cooling they may be moved to dry cask storage. Used fuel is unloaded into a storage pond immediately adjacent to the reactor to allow the radiation levels to decrease. In the ponds, the water shields the radiation and absorbs the heat. Used fuel is held in such pools for several months to several years. Depending on policies in particular countries, some used fuel may be transferred to central storage facilities. There are two approaches adopted regarding spent or used fuel. Countries like USA have adopted an open cycle for the civil nuclear plants, where the spent fuel after sufficient cooling and activity decrease is treated as a waste. In countries with scant nuclear fuel resources, a closed cycle is adopted, in which the spent fuel after some cooling is reprocessed and useful unutilized uranium and plutonium taken out. The remaining waste has a much lower level of activity and can be disposed suitably.

4.10 REPROCESSING

Used fuel is about 95% uranium-238 but it also contains up to 1% uranium-235 that has not fissioned, about 1% plutonium and 3% fission products, which are highly radioactive. In a reprocessing facility the used fuel is separated into its three components: uranium, plutonium and waste containing fission products. Reprocessing enables recycling of the uranium and plutonium into fresh fuel, and produces a significantly reduced amount of waste (compared with treating all used fuel as waste). Spent fuel discharged from reactors contains appreciable quantities of fissile (U-235 and Pu-239), fertile (U-238), and other radioactive materials, including neutron poisons. These fissile and fertile materials can be chemically separated and recovered from the spent fuel. The recovered uranium and plutonium can be recycled for use as nuclear fuel. The reprocessed uranium, which constitutes the bulk of the spent fuel material, can in principle also be re-used as fuel. Finally, the breeder reactor can employ not only the recycled plutonium and uranium in spent fuel, but all the actinides, closing the nuclear fuel cycle and potentially multiplying the energy extracted from natural uranium by more than 60 times. Nuclear reprocessing also reduces the volume of high-level nuclear waste and its radio toxicity.

Despite the energy and waste disposal benefits obtainable through nuclear reprocessing, reprocessing has been politically controversial because of the potential to contribute to nuclear proliferation, the potential vulnerability to nuclear terrorism, and because of its high cost compared to the once-through fuel cycle.

Currently, plants in Europe are reprocessing spent fuel from utilities in Europe and Japan. Reprocessing of spent commercial-reactor nuclear fuel is currently not permitted in the United States due to the perceived danger of nuclear proliferation. However the recently announced Global Nuclear Energy Partnership would see the U.S. form an international partnership to see spent nuclear fuel reprocessed in a way that renders the plutonium in it usable for nuclear fuel but not for nuclear weapons. Two techniques used in reprocessing are solvent extraction and pyroprocessing.

4.10.1 Solvent Extraction

4.10.1.1 PUREX

PUREX, the current standard method, is an acronym standing for **P**lutonium and **U**ranium **R**ecovery by **EX**traction. The PUREX process is a liquid-liquid extraction method used to reprocess spent nuclear fuel, in order to extract uranium and plutonium, independent of each other, from the fission products. Tri Butyl Phosphate (TBP) is generally used as the extractant. This is the most developed and widely used process in the industry at present. When used on fuel from commercial power reactors the plutonium extracted typically contains too much Pu-240 to be useful in a nuclear weapon. However, reactors that are capable of refueling frequently can be used to produce weapon-grade plutonium, which can later be recovered using PUREX.

PUREX Reprocessing of Spent Fuel

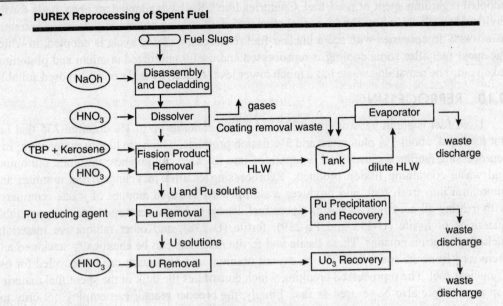

Fig. 4.10: Schematic of PUREX process

4.10.1.2 UREX

The PUREX process can be modified to make a UREX process which could be used to save space inside high level nuclear waste disposal sites, such as the Yucca Mountain nuclear waste repository, by removing the uranium which makes up the vast majority of the mass and volume of used fuel and recycling it as reprocessed uranium.

The UREX (**UR**anium **EX**traction) process is a PUREX process which has been modified to prevent the plutonium from being extracted. This can be done by adding a plutonium reductant before the first metal extraction step. In the UREX process, ~99.9% of the Uranium and >95% of Technetium are separated from each other and the other fission products and actinides. The key is the addition of acetohydroxamic acid (AHA) to the extraction and scrub sections of the process. The addition of AHA greatly diminishes the extractability of Plutonium and Neptunium, providing greater proliferation resistance than with the plutonium extraction stage of the PUREX process. Thus uranium which forms a large part of spent fuel is recovered and waste to be handled is only Pu.

4.10.1.3 TRUEX

Adding a second extraction agent, octyl(phenyl)-N, N-dibutyl carbamoylmethyl phosphine oxide (CMPO) in combination with tributylphosphate (TBP), the PUREX process can be turned into the TRUEX (**TR**ans**U**ranic **EX**traction) process. TRUEX was invented in the USA by Argonne National Laboratory and is designed to remove the transuranic metals (Am/Cm) from waste. The idea is that by lowering the alpha activity of the waste, the majority of the waste can then be disposed of with greater ease. Like PUREX, this is a solvent extraction process.

4.10.1.4 DIAMEX

As an alternative to TRUEX, an extraction process using a malondiamide has been devised. The DIAMEX (**DIAM**ide**EX**traction) process has the advantage of avoiding the formation of organic waste which contains elements other than Carbon, Hydrogen, Nitrogen, and Oxygen. Such an organic waste can be burned without the formation of acidic gases which could

Reactor Fuel Cycle

contribute to acid rain. The DIAMEX process is being worked on in Europe by the French CEA. The process is sufficiently mature that an industrial plant could be constructed with the existing knowledge of the process. In common with PUREX this is a solvent extraction process.

4.10.2 Pyroprocessing

Pyroprocessing is a generic term for high-temperature methods. Solvents are molten salts (e.g. LiCl + KCl or LiF + CaF_2) and molten metals (e.g. cadmium, bismuth, magnesium) rather than water and organic compounds. Electro-refining, distillation, and solvent-solvent extraction are common steps.

4.10.2.1 Electrolysis

These processes were developed by Argonne National Laboratory and used in the Integral Fast Reactor project. PYRO-A is a means of separating actinides (elements within the actinide family, generally heavier than U-235) from non-actinides. The spent fuel is placed in an anode basket which is immersed in a molten salt electrolyte. An electrical current is applied, causing the uranium metal (or sometimes oxide, depending on the spent fuel) to plate out on a solid metal cathode while the other actinides (and the rare earths) can be absorbed into a liquid cadmium cathode. Many of the fission products (such as caesium, zirconium and strontium) remain in the salt. As alternatives to the molten cadmium electrode it is possible to use a molten bismuth cathode, or a solid aluminium cathode. (Fig. 4.11)

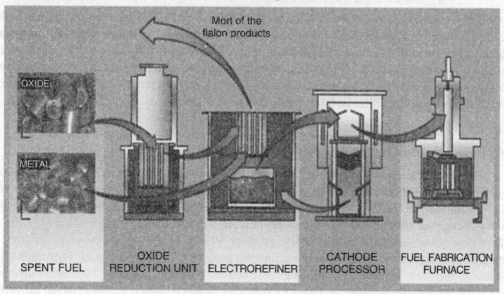

Fig. 4.11: Pyro Processing

Since the majority of the long term radioactivity, and volume, of spent fuel comes from actinides, removing the actinides produces waste that is more compact, and not nearly as dangerous over the long term. The radioactivity of this waste will then drop to the level of various naturally occurring minerals and ores within a few hundred, rather than thousands of years.

The mixed actinides produced by pyrometallic processing can be used again as nuclear fuel, as they are virtually all either fissile, or fertile, though many of these materials would require a fast breeder reactor in order to be burned efficiently. In a thermal neutron spectrum, the concentrations of several heavy actinides (curium-242 and plutonium-240) can become quite high, creating fuel that is substantially different from the usual uranium or mixed uranium-plutonium oxides (MOX) that most current reactors were designed to use.

Another pyrochemical process, the PYRO-B process, has been developed for the processing and recycling of fuel from a transmuter reactor (a fast breeder reactor designed to convert transuranic nuclear waste into fission products). A typical transmuter fuel is free from uranium and contains recovered transuranics in an inert matrix such as metallic zirconium. In the PYRO-B processing of such fuel, an electro refining step is used to separate the residual transuranic elements from the fission products and recycle the transuranics to the reactor for fissioning. Newly-generated technetium and iodine are extracted for incorporation into transmutation targets, and the other fission products are sent to waste.

4.11 WASTE MANAGEMENT

Nuclear waste management is a critical part of the fuel cycle. Wastes from the nuclear fuel cycle are categorized as high-, medium- or low-level wastes by the amount of radiation that they emit. These wastes come from a number of sources and include:

- Low-level waste produced at all stages of the fuel cycle.
- Intermediate-level waste produced during reactor operation and by reprocessing.
- High-level waste, which is waste containing fission products from reprocessing, and in many countries, the used fuel itself.

The enrichment process leads to the production of much 'depleted' uranium, in which the concentration of uranium-235 is significantly less than the 0.7% found in natural uranium. Small quantities of this material, which is primarily uranium-238, are used in applications where high-density material is required, including radiation shielding and in the production of MOX fuel. While 238U is not fissile, it is a low specific activity radioactive material and some precautions must, therefore, be taken in its storage or disposal.

4.11.1 Types of Radioactive Wastes

4.11.1.1 Exempt waste & very low level waste

Exempt waste and very low level waste (VLLW) contains radioactive materials at a level which is not considered harmful to people or the surrounding environment. It consists mainly of broken material (such as concrete, plaster, bricks, metal, valves, piping etc) produced during rehabilitation or dismantling operations on nuclear industrial sites. Other industries, such as food processing, chemical, steel etc also produce VLLW as a result of the concentration of natural radioactivity present in certain minerals used in their manufacturing processes. The waste is therefore disposed of with domestic refuse, although countries such as France are currently developing facilities to store VLLW in specifically designed VLLW disposal facilities.

4.11.1.2 Low-level waste

Low-level waste (LLW) is generated from hospitals and industry, as well as the nuclear fuel cycle. It comprises paper, rags, tools, clothing, and filters etc, which contain small amounts of mostly short-lived radioactivity. It does not require shielding during handling and transport and is suitable for shallow land burial. To reduce its volume, it is often compacted or incinerated before disposal. It comprises some 90% of the volume but only 1% of the radioactivity of all radioactive waste.

4.11.1.3 Intermediate-level waste

Intermediate-level waste (ILW) contains higher amounts of radioactivity and requires some shielding. It typically comprises resins, chemical sludge and metal fuel cladding, as well as contaminated materials from reactor decommissioning. Smaller items and any non-solids may be solidified in concrete or bitumen for disposal. It makes up some 7% of the volume and has 4% of the radioactivity of all radwaste.

4.11.1.4 High-level waste

High-level waste (HLW) arises from the 'burning' of uranium fuel in a nuclear reactor. HLW contains the fission products and transuranic elements generated in the reactor core. It is highly radioactive and hot, so requires cooling and shielding. It can be considered as the 'ash' from 'burning' uranium. HLW accounts for over 95% of the total radioactivity produced in the process of electricity generation. There are two distinct kinds of HLW:

- Used fuel itself.
- Separated waste from reprocessing the used fuel. HLW has both long-lived and short-lived components, depending on the length of time it will take for the radioactivity of particular radio nuclides to decrease to levels that are considered no longer hazardous for people and the surrounding environment. If generally short-lived fission products can be separated from long-lived actinides, this distinction becomes important in management and disposal of HLW.

4.11.2 Treatment and Conditioning of Nuclear Wastes

Treatment and conditioning processes are used to convert radioactive waste materials into a form that is suitable for its subsequent management, such as transportation, storage and final disposal. The principal aims are to:

- Minimise the volume of waste requiring management via treatment processes.
- Reduce the potential hazard of the waste by conditioning it into a stable solid form that immobilises it and provides containment to ensure that the waste can be safely handled during transportation, storage and final disposal.

It is important to note that, while treatment processes such as compaction and incineration reduce the volume of waste, the amount of radioactivity remains the same. As such, the radioactivity of the waste will become more concentrated as the volume is reduced.

Conditioning processes such as cementation and vitrification are used to convert waste into a stable solid form that is insoluble and will prevent dispersion to the surrounding environment. A systematic approach incorporates:

- Identifying a suitable matrix material - such as cement, bitumen, polymers or borosilicate glass - that will ensure stability of the radioactive materials for the period necessary. The type of waste being conditioned determines the choice of matrix material and packaging.
- Immobilising the waste through mixing with the matrix material.
- Packaging the immobilised waste in, for example, metallic drums, metallic or concrete boxes or containers, copper canisters.

The choice of process used is dependent on the level of activity and the type (classification) of waste. Each country's nuclear waste management policy and its national regulations also influence the approach taken.

4.11.2.1 Incineration

Incineration of combustible wastes can be applied to both radioactive and other wastes. In the case of radioactive waste, it has been used for the treatment of low-level waste from nuclear power plants, fuel production facilities, research centres (such as biomedical research), medical sector and waste treatment facilities.

Following the segregation of combustible waste from non-combustible constituents, the waste is incinerated in a specially engineered kiln up to around 1000°C. Any gases produced during incineration are treated and filtered prior to emission into the atmosphere and must conform to international standards and national emissions regulations.

Following incineration, the resulting ash, which contains the radionuclides, may require further conditioning prior to disposal such as cementation or bituminisation. Compaction technology may also be used to further reduce the volume, if this is cost-effective. Volume reduction factors of up to around 100 are achieved, depending on the density of the waste.

Incineration technology is subject to public concern in many countries as local residents worry about what is being emitted into the atmosphere. However, modern incineration systems are well engineered, high technology processes designed to completely and efficiently burn the waste whilst producing minimum emissions.

The incineration of hazardous waste (e.g. waste oils, solvents) and non-hazardous waste (municipal waste, biomass, tyres, sewage sludge) is also practised in many countries.

4.11.2.2 Compaction

Compaction is a mature, well-developed and reliable volume reduction technology that is used for processing mainly solid man-made low-level waste (LLW). Some countries (Germany, UK and USA) also use the technology for the volume reduction of man-made intermediate-level/transuranic waste. Compactors can range from low-force compaction systems (~5 tonnes or more) through to presses with a compaction force over 1000 tonnes, referred to as supercompactors. Volume reduction factors are typically between 3 and 10, depending on the waste material being treated.

Low-force compaction is typically applied to the compression of bags of rubbish, in order to facilitate packaging for transport either to a waste treatment facility, where further compaction might be carried out, or to a storage/disposal facility. In the case of supercompactors, in some applications, waste is sorted into combustible and non-combustible materials. Combustible waste is then incinerated whilst non-combustible waste is supercompacted. In certain cases, incinerator ashes are also supercompacted in order to achieve the maximum volume reduction.

Low-force compaction utilises a hydraulic or pneumatic press to compress waste into a suitable container, such as a 200-litre drum. In the case of a supercompactor, a large hydraulic press crushes the drum itself or other receptacle containing various forms of solid low- or intermediate-level waste (LLW or ILW). The drum or container is held in a mold during the compaction stroke of the supercompactor, which minimises the drum or container outer dimensions. The compressed drum is then stripped from the mold and the process is repeated. Two or more crushed drums, also referred to as pellets, are then sealed inside an overpack container for interim storage and/or final disposal.

4.11.2.3 Cementation

Cementation through the use of specially formulated grouts provides the means to immobilise radioactive material that is on solids and in various forms of sludges and precipitates/gels (flocks) or activated materials.

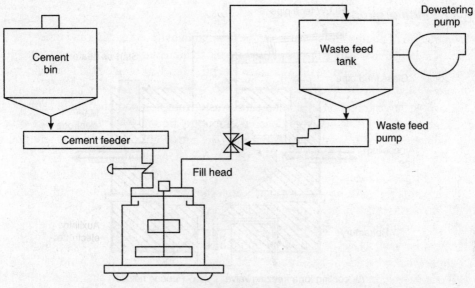

Fig. 4.12: Cementation

In general the solid wastes are placed into containers (Figure 4.12). The grout is then added into this container and allowed to set. The container with the now monolithic block of concrete/waste is then suitable for storage and disposal. Similarly in the case of sludges and flocks, the waste is placed in a container and the grouting mix, in powder form, is added. The two are mixed inside the container and left to set leaving a similar type of product as in the case of solids, which can be disposed of in a similar way.

This process has been used for example in small oil drums and 500-litre containers for intermediate-level wastes and has been extended to ISO shipping containers for low-level waste materials.

The technology is being used in the immobilisation of many toxic and hazardous wastes that arise outside the nuclear industry and has the potential to be used in many more cases.

4.11.2.4 Vitrification

The immobilisation of high-level waste (HLW) requires the formation of an insoluble, solid waste form that will remain stable for many thousands of years. In general borosilicate glass has been chosen as the medium for dealing with HLW. The stability of ancient glass for thousands of years highlights the suitability of borosilicate glass as a matrix material.

This type of process, referred to as vitrification, has also been extended for lower level wastes where the type of waste or the economics have been appropriate. Most high-level wastes other than spent fuel itself, arise in a liquid form from the reprocessing of spent fuel. To allow incorporation into the glass matrix this waste is initially calcined (dried) which turns it into a solid form. This product is then incorporated into molten glass in a stainless container and allowed to cool, giving a solid matrix (Figure 4.13). The containers are then welded closed and are ready for storage and final disposal. This process is currently being used in France, Japan, the former Soviet Union, UK and USA and is seen as the preferred process for management of separated HLW arising from reprocessing.

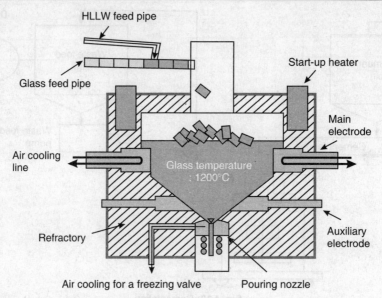

Fig. 4.13: Vitrification

4.12 WASTE DISPOSAL METHODS
4.12.1 Near-surface Disposal

The International Atomic Energy Agency (IAEA) definition of this option is the disposal of waste, with or without engineered barriers, in:

- Near-surface disposal facilities at ground level. These facilities are on or below the surface where the protective covering is of the order of a few metres thick. Waste containers are placed in constructed vaults and when full the vaults are backfilled. Eventually they will be covered and capped with an impermeable membrane and topsoil. These facilities may incorporate some form of drainage and possibly a gas venting system.

Near-surface disposal facilities are in caverns below ground level. Unlike near-surface disposal at ground level where the excavations are conducted from the surface, shallow disposal requires underground excavation of caverns but the facility is at a depth of several tens of metres below the Earth's surface.

The term near-surface disposal replaces the terms 'shallow land' and 'ground disposal', but these older terms are still sometimes used when referring to this option. These facilities will be affected by long-term climate changes (such as glaciation) and this effect must be taken into account when considering safety as such changes could cause disruption of these facilities. This type of facility is therefore typically used for LLW and ILW with a radionuclide content of short half-life (up to about 30 years).

4.12.2 Deep Geological Disposal

The long timescales over which some of the waste remains radioactive led to the idea of deep geological disposal in underground repositories in stable geological formations. Isolation is provided by a combination of engineered and natural barriers (rock, salt, clay) and no obligation to actively maintain the facility is passed on to future generations. This is often termed a multi-barrier concept, with the waste packaging, the engineered repository and the geology all providing barriers to prevent the radionuclides from reaching humans and the environment.

Table 4.1: WASTE DISPOSAL FACILITIES

Option	Examples
• Near Surface Disposal at ground level, or in caverns below ground level (at depths of tens of metres)	• Implemented for LLW in many countries, including Czech Republic, Finland, France, Japan, Netherlands, Spain, Sweden, UK and USA. • Implemented in Finland and Sweden for LLW and short-lived ILW.
• Deep Geological Disposal (at depths between 250m and 100 m)	• Most countries with high-level and long-lived radioactive waste have investigated deep geological disposal and it is official policy in various countries (variations also include multinational facilities). • Implemented in USA for defence-related ILW. • Preferred sites for HLW/spent fuel selected in France, Sweden, Finland and USA. • Geological repository site selection process commenced in UK and Canada.

A repository is comprised of mined tunnels or caverns into which packaged waste would be placed. In some cases (e.g. wet rock) the waste containers are then surrounded by a material such as cement or clay (usually bentonite) to provide another barrier (called buffer and/or backfill). The choice of waste container materials and design and buffer/backfill material varies depending on the type of waste to be contained and the nature of the host rock-type available.

Deep geological disposal remains the preferred option for waste management of long-lived radioactive waste in several countries, including Argentina, Australia, Belgium, Czech Republic, Finland, Japan, Netherlands, Republic of Korea, Russia, Spain, Sweden, Switzerland and USA. Hence, there is much information available on different disposal concepts; a few examples are given here. The only purpose-built deep geological repository for long-lived ILW that is currently licensed for disposal operations is in the USA. Plans for disposal of spent fuel are well advanced in Finland, Sweden and the USA. In Canada and the UK, deep disposal has been selected and the site selection process has commenced. (See table 4.1)

The Swedish proposed KBS-3 disposal concept uses a copper container with a steel insert to contain the spent fuel. After placement in the repository about 500 metres deep in the bedrock, the container would be surrounded by a bentonite clay buffer to provide a very high level of containment of the radioactivity in the wastes over a very long time period. In June 2009, the Swedish Nuclear Fuel and Waste Management Company (SKB) announced its decision to locate the repository at Östhammar (Forsmark).

Finland's repository programme is also based on the KBS-3 concept. Spent nuclear fuel packed in copper canisters will be embedded in the Olkiluoto bedrock at a depth of around 400 metres. The country's nuclear waste management company, Posiva Oy, expects the repository to begin disposal operations in 2020.

The deposits of native (pure) copper in the world have proven that the copper also used in the final disposal container can remain unchanged inside the bedrock for extremely long periods, if the geochemical conditions are appropriate (reducing groundwaters). The findings of ancient copper tools, many thousands of years old, also demonstrate the long-term corrosion resistance of copper, making it a credible container material for long-term radioactive waste storage.

4.12.3 Disposal in Outer Space

The objective of this option is to remove the radioactive waste from the Earth, for all time, by ejecting it into outer space. The waste would be packaged so that it would be likely to remain intact under most conceivable accident scenarios. A rocket or space shuttle would be used to launch the packaged waste into. Because of the high cost of this option and the safety aspects associated with the risk of launch failure, this option has not found favour.

4.12.4 Deep Boreholes

For the deep borehole option, solid packaged wastes would be placed in deep boreholes drilled from the surface to depths of several kilometres with diameters of typically less than 1 metre. The waste containers would be separated from each other by a layer of bentonite or cement. The borehole would not be completely filled with wastes. The top two kilometres would be sealed with materials such as bentonite, asphalt or concrete.

Deep borehole concepts have been developed (but not implemented) in several countries, including Denmark, Sweden, Switzerland and USA for HLW and spent fuel. Compared with deep geological disposal in a mined underground repository, placement in deep boreholes is considered to be more expensive for large volumes of waste. This option was abandoned in countries such as Finland and USA. The feasibility of disposal of spent fuel in deep boreholes has been studied in Sweden, in order to check whether deep geological disposal remains the preferred option. The borehole concept remains an attractive proposition under investigation for the disposal of sealed radioactive sources from medical and industrial applications.

4.12.5 Disposal at Sea

Disposal at sea involves radioactive waste being shipped out to sea and dropped into the sea in packaging designed to either: implode at depth, resulting in direct release and dispersion of radioactive material into the sea; or sink to the seabed intact. Over time the physical containment of containers would fail, and radionuclides would be dispersed and diluted in the sea. Further dilution would occur as the radionuclides migrated from the disposal site, carried by currents. The amount of radionuclides remaining in the sea water would be further reduced both by natural radioactive decay, and by the removal of radionuclides to seabed sediments by the process of sorption. This method is not permitted by a number of international agreements.

The application of the sea disposal of LLW and ILW has evolved over time from being a disposal method that was actually implemented by a number of countries, to one that is now banned by international agreements. Countries that have at one time or another undertaken sea disposal using the above techniques include Belgium, France, Federal Republic of Germany, Italy, Netherlands, Sweden, Switzerland and the UK, as well as Japan, South Korea, and the USA. This option has not been implemented for HLW.

SUMMARY

This chapter took the reader through the different routes the fuel takes right from exploration, mining, fabrication, use in a reactor to reprocessing and waste management. Various methods of reprocessing have been outlined, besides the types of waste treatment and storage.

BIBLIOGRAPHY

1. P.A.Baisder and G.R.Choppin, Radiochemistry and Nuclear Chemistry.
2. P.D.Wilson, The Nuclear Fuel Cycle: From Ore to Waste, Oxford Science Publications (1996).
3. G.Robert, N. Cochran, Tsoulfanidis, Robert G.Cochran, W.F.Miller, Nuclear Fuel Cycle : Analysis and Management, American Nuclear Society, 2nd Edidtion, (1993).
4. James Saling, Radioactive Waste Management, CRC Press, 2nd Edition (2001).
5. R.L.Murray, Understanding Radioactive Waste, Battelle Press; 5th edition, (2003).
6. Richard Stephenson, Introduction to Nuclear Engineering, McGraw Hill, 2nd Edition, (1958).
7. Kenneth D. Kok (ed.), Handbook of Nuclear Engineering, CRC Press, (2009).
8. Dan Gabriel Cacuci (ed.), Handbook of Nuclear Engineering, Springer (2010).
9. Steven B., Krivit, Jay H.Lehr, Thomas B. Kingery, Nuclear Energy Encyclopedia: Science, Technology, and Applications, John Wiley (2011).
10. K.Raj, K.K.Prasad,N.K.Bansal, Radioactive waste Management Practices in India, Nuclear Engineering and Design, 236, (2006) 914-930.
11. P.K.Dey, N.K.Bansal, Spent Fuel Reprocessing: A Vital link to Indian nuclear power program, Nuclear Engineering and Design, 236 (2006) 723-729.

ASSIGNMENTS

1. Explain with a schematic the components of a nuclear fuel cycle.
2. What do you mean by front end and back end of the fuel cycle?
3. Give the material balance in the nuclear fuel cycle for a 1000 MWe nuclear power plant.
4. What do you mean by (*a*) Uranium Mining (*b*) Uranium Milling.
5. What is meant by enrichment? How is Uranium fuel enriched?
6. What is the need for interim storage of spent fuel?
7. Why is the spent fuel needed to be reprocessed? What are the different steps in fuel reprocessing?
8. Explain briefly the different processes of reprocessing.
9. How are nuclear wastes classified?
10. Explain briefly the different steps in radioactive waste management.

5

Components of a Nuclear Reactor

5.0 INTRODUCTION

Earlier chapters have introduced the students into the basic physics of nuclear reactors where energy is released in fission. The heat energy produced during fission is used to generate steam which runs a turbine generator. The path taken by the steam and related thermodynamic processes is the subject of Chapter 3. Before getting into the details of different reactor systems, it is essential to know the different components that go to make a nuclear power plant. This is the subject of this chapter.

5.1 FERMI PILE

It would be worthwhile here to have a look at the first reactor in which Enrico Fermi and colleagues, proved the possibility of a sustained fission reaction (Fig. 5.1). The date was December 2nd, 1942, and for the very first time, man created a fission chain reaction. The credit for this achievement goes to a Chicago team led by Enrico Fermi (1901–1954). On that day, the neutron population scattering in the pile amplified very gradually, even after the neutron source was withdrawn. When the nuclear power level reached about half a watt, the cautious Fermi ordered the insertion of the cadmium control rod to stop the criticality. The event was immortalised by a table and a drawing, reproduced below (see Figure 5.1).

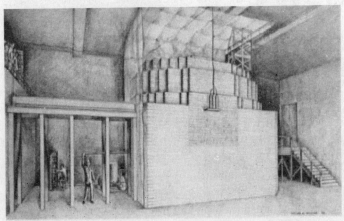

Fig. 5.1: FERMI PILE (Courtesy Argonne Nat. Lab, Chicago)

Components of a Nuclear Reactor

They demonstrated that the critical condition (the configuration allowing the chain reaction to be self-sustaining) was reached when 400 tonnes of graphite, 6 tonnes of uranium metal and 37 tonnes of uranium oxide were piled up in a carefully planned arrangement (this explains the origin of the term atomic pile, which we often use to refer to a nuclear reactor). Some of the main principles later to be applied in all reactors, both research reactors and power plants, were already used in Fermi's pile.

5.1.1 Control

This was symbolised by the two operators at the bottom: on the left, the operator monitoring the detector display represents the monitoring function. On the right, the operator in charge of the cadmium control rod represents the control function. Cadmium is an efficient neutron-capturing material. When the rod is pushed in, the number of neutrons captured by the cadmium increases. This reduces the number of neutrons causing fission in the uranium. The chain reaction is then stifled. Conversely, if the rod is pulled out slightly, more neutrons become available to cause fission reactions. The chain reaction is then amplified. To control the system according to requirements, the monitoring and control functions must talk to each other (in this case, simply a verbal dialogue between the two operators).

5.1.2 Safety

Safety depends first and foremost on good monitoring and control. It also requires an emergency stop mechanism in the event of an incident. In this experiment, the emergency stop function is provided by an unseen operator located above the pile. This person is armed with an axe, and on Fermi's signal can cut the rope holding an emergency cadmium control rod. The last line of defense consisted in a tank of cadmium salt solution to release the solution into the pile.

5.1.3 Radiation Monitoring

Shielding is provided in this case by a detector hanging in front of the pile to measure the ambient radiation level. The signal passes through the cable running along the ceiling to a display placed in view of Fermi himself, on the balcony. Fermi can thereby ensure that he and his colleagues do not run the risk of excessive irradiation and can trigger the emergency stop if necessary. The criticality of Fermi's pile concluded half a century of very active research in nuclear physics.

5.2 REACTOR CORE

The core of the reactor contains all of the nuclear fuel and generates all of the heat. It contains low-enriched uranium (<5% U-235), control systems, and structural materials. The core can contain hundreds of thousands of individual fuel pins (Fig 5.2).

The smallest unit of the reactor is the fuel pin. These are typically uranium-oxide (UO_2), but can take on other forms, including thorium-bearing material. They are often surrounded by a metal tube (called the cladding) to keep fission products from escaping into the coolant. Fuel assemblies are bundles of fuel pins (Fig 5.2). Fuel is put in and taken out of the reactor in assemblies. The assemblies have some structural material to keep the pins close but not touching, so that there's room for coolant.

This is a full core (Fig 5.3), made up of several hundred assemblies. Some assemblies are control assemblies. Various fuel assemblies around the core have different fuel in them. They vary in enrichment and age, among other parameters. The assemblies may also vary with height, with different enrichments at the top of the core from those at the bottom.

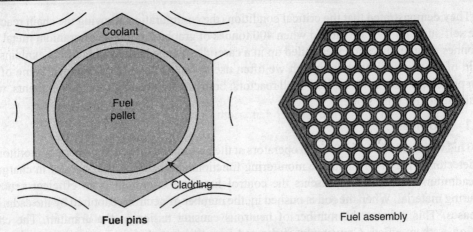

Fig. 5.2: Fuel Pin and Fuel assembly

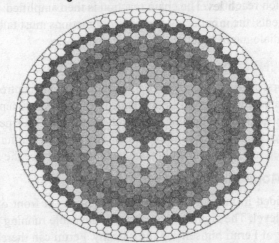

Fig 5.3: Full core

5.3 COOLANT

A coolant is necessary to absorb and remove the heat produced by nuclear fission and maintain the temperature of the fuel within acceptable limits. It can then transfer the heat to drive electricity-generating turbines. If water is used as the coolant the steam produced can be fed directly to the turbines. Alternatively, it can be passed through a heat exchanger which will remove the heat and produce the necessary steam. Other possible coolants are heavy water, gases like carbon dioxide or helium, or molten metals such as sodium or lead and bismuth. A coolant can also be a moderator; water is used in this dual way in most modern reactors.

5.4 CONTROL RODS

Control rods are made of materials that absorb neutrons, for example, boron, silver, indium, cadmium and hafnium. They are introduced into the reactor to reduce the number of neutrons and thus stop the fission process when required; or, during operation, to control and regulate the level and spatial distribution of power in the reactor.

Components of a Nuclear Reactor

5.5 MODERATOR

A moderator is necessary to slow down the fast neutrons created during fission to the thermal energy range so as to increase their efficiency in causing further fission. The moderator must be a light material that will allow the neutrons to slow down without being captured. Usually, ordinary water is used; alternatives in use are graphite, a form of carbon, and heavy water, which is formed with the heavier deuterium isotope of hydrogen.

5.6 OTHER CORE COMPONENTS

The fuel along with the mechanical structure that holds it in place, forms the reactor core. Typically, a neutron reflector surrounds the core and serves to return as many of the neutrons as possible that have leaked out of the core and therefore maximize the efficiency of their use. Often, the coolant and/or moderator serve as the reflector. The core and reflector are often housed in a thick steel container called the reactor pressure vessel (Fig. 5.4). Radiation shielding is provided to reduce the high levels of radiation produced by the fission process. Numerous instruments are inserted into the core and support systems to permit the monitoring and control of the reactor.

5.7 CONTAINMENT (FIG 5.4)

All the components of the reactor are contained in a solid concrete structure that guarantees further isolation from external environment. This structure is made of concrete that is one-metre thick, covered by steel. The most recent reactors sometimes contain two containment structures and are designed to defy all types of accidental events, even the impact of an aircraft.

5.8 CORE CATCHER

This is provided below the core to collect molten fuel in case of an accidental situation. The shape is such that it moves the molten fuel away from the centre of the core and does not allow a critical geometry to the molten fuel.

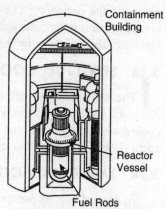

Fig. 5.4: Containment Building

5.9 STEAM GENERATOR

The steam generator utilizes the heat of the core transported by the coolant to convert high pressure water into steam. As we shall see in later chapters, there is no separate steam generator in Boiling Water Reactor, as steam is produced in the core itself. In the steam generator the high pressure fluid is generally on the tube side and the low pressure fluid on the shell side. This is based on the following considerations:

-Thickness of a tube (for a cylinder) based on pressure is given by

$$\sigma_\theta = \frac{Pr}{t}$$

where
- P is the internal pressure
- t is the wall thickness

- r is the inside radius of the cylinder.
- σ_θ is the hoop stress.

If the higher pressure fluid is on the shell side, it calls for a higher thickness of shell for a given limit on hoop stress as thickness is proportional to radius. This increases the cost of the steam generator/heat exchanger.

5.10 TURBINE GENERATOR

The steam from the steam generator drives the turbine, which is connected to an electrical generator for producing electrical power.

5.11 STEAM/WATER SYSTEM

This is similar to a conventional fossil fired power station, comprising feed water heaters, condenser, boiler feed pumps, cooling tower etc.

5.12 FUEL HANDLING

The fuel gets consumed as the reactor operates. The fuel at the centre of the core sees a higher neutron flux and those at the periphery see less and less. So after the central fuel assembly and say the first ring of assemblies see the planned burnup, these are removed and placed in the periphery. The assemblies in the periphery are brought to the centre. The fuel management is to be done such that all assemblies see the maximum burnup. This calls for fuel handling which is done by the fuel handling machines.

5.13 SPENT FUEL COOLING

The fuel assemblies are taken out of the reactor vessel after reaching the targeted burnup. They still have heat production of the order of few watts or Kilowatts. Hence they need to be cooled when they are stored outside prior to dispatch for reprocessing. This is referred to as spent (Burnt) fuel handling. In most reactors the spent fuel is kept in water bays.

5.14 EMERGENCY CORE COOLING

In case of loss of pumping power the core needs to be cooled. For this purpose, storage tanks with enough water combined with diesel driven pumps are used. These and similar type of cooling are referred to as Emergency core cooling devices.

5.15 TYPES OF NUCLEAR REACTORS

There are very many different types of nuclear reactors with different fuels, coolants, fuel cycles and purposes. The evolution of the different types has evolved in different countries. Table 5.1 lists the different types and number of reactors built and operated in the world. However all the designs are having the same degree of safety and are based on the guidelines of the International Atomic Energy Agency (IAEA).

5.15.1 Pressurised Water Reactor(PWR)

The largest number of reactors is PWR. A schematic is given in Fig. 5.5. The coolant and moderator are both light water. The coolant picks up the heat from the core and exchanges it to another light water circuit in a steam generator, to produce steam for running a turbine generator. It uses U235 enriched to about 5% as compared to natural uranium which contains only 0.7% U235.

Components of a Nuclear Reactor

Table 5.1: Nuclear Power Plants

Reactor Type	Number	GWe	Fuel	Coolant	Moderator
PWR	271	270.4	Enriched UO_2	Water	Water
BWR	84	81.2	Enriched UO_2	Water	Water
CANDU	48	27.1	Natural UO_2	Heavy Water	Heavy Water
GCR	17	9.6	Enriched UO_2	CO_2	Graphite
RBMK	11	10.4	Enriched UO_2	Water	Graphite
SFR	6	0.6	PuO_2, UO_2	Sodium	NIL

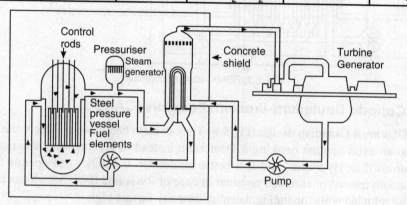

Figure 5.5: Pressurised Water Reactor Schematic

5.15.2 Boiling Water Reactor (BWR)

Second most common, the BWR is similar to the PWR in many ways. However, they only have one coolant loop. The water boils in the reactor core and produces saturated steam, which goes out at the top of the reactor to the turbine (Fig 5.6). Since the steam produced in the core is used to drive the turbine any activity picked up by the coolant will reach the turbine, thereby making maintenance of steam/water systems slightly difficult. Since there is only one coolant circuit, in place of two in a PWR, the capital cost is less for BWR. Large amount of operating experience has been accumulated and the designs and procedures have been largely optimized.

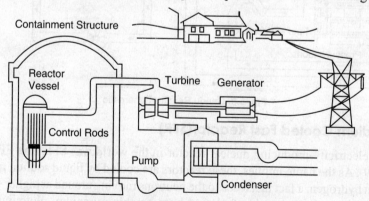

Fig. 5.6: Boiling Water Reactor Schematic

Unlike the BWR with a pressure vessel, the Russians followed a design with pressure tubes and replaced the light water moderator with solid graphite blocks. This is the type of reactor involved in the Chernobyl accident. Fig 5.7 shows a schematic of the same.

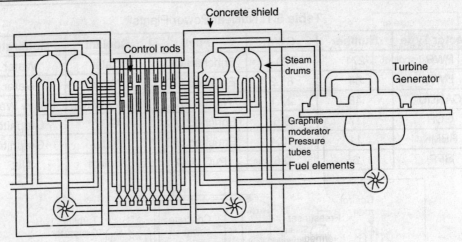

Fig. 5.7: RBMK reactor Schematic

5.15.3 Canada Deuterium-Uranium Reactors (CANDU)

CANDUs are a Canadian design (Fig 5.8) They contain heavy water, where the Hydrogen in H2O has an extra neutron (making it Deuterium instead of Hydrogen). Deuterium absorbs fewer neutrons than Hydrogen, and is a better moderator. CANDUs can operate using only natural uranium instead of enriched uranium in case of PWR and BWR which use light water. They can be refueled while operating, keeping capacity factors high.

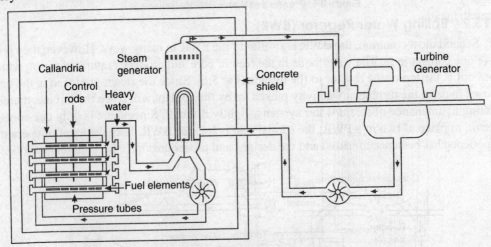

Fig. 5.8: Candu Reactor Schematic

5.15.4 Sodium Cooled Fast Reactor(SFR)

The first electricity-producing nuclear reactor in the world was SFR (the EBR-1 in Arco, Idaho)(Fig 5.9). As the name implies, these reactors are cooled by liquid sodium metal. Sodium is heavier than hydrogen, a fact that leads to the neutrons moving around at higher speeds (hence *fast*). These can use metal or oxide fuel, and burn anything uranium, plutonium and higher actinides. They can breed its own fuel, effectively eliminating any concerns about uranium shortages Sodium coolant is however reactive with air and water. Thus, leaks of sodium need to be minimized by suitable design.

Components of a Nuclear Reactor

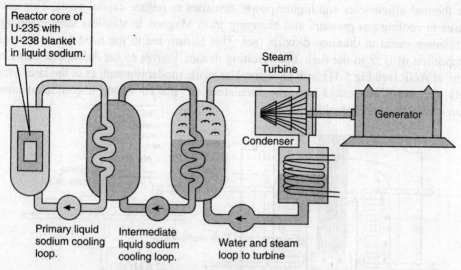

Fig. 5.9: Sodium Cooled Fast Reactor Schematic

5.15.5 Advanced Gas Cooled Reactors (AGR)

Of the main commercial reactor types, two (Magnox and AGR) owe much to the very earliest reactor designs in that they are graphite moderated and gas cooled. Magnox reactors (see Fig .5.10) were built in the UK from 1956 to 1971 but have now been superseded. The Magnox reactor is named after the magnesium alloy used to encase the fuel, which is natural uranium metal.

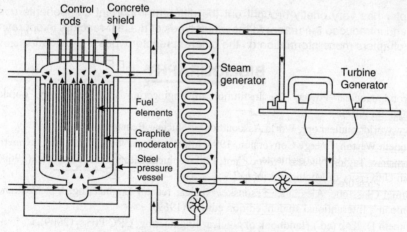

Fig. 5.10: Magnox Reactor

Fuel elements consisting of fuel rods encased in Magnox cans are loaded into vertical channels in a core constructed of graphite blocks. Further vertical channels contain control rods (strong neutron absorbers) which can be inserted or withdrawn from the core to adjust the rate of the fission process and, therefore, the heat output. The whole assembly is cooled by blowing carbon dioxide gas past the fuel cans, which are specially designed to enhance heat transfer. The hot gas then converts water to steam in a steam generator. Early designs used a steel pressure vessel, which was surrounded by a thick concrete radiation shield. In later designs, a dual-purpose concrete pressure vessel and radiation shield was used. In order to improve the cost effectiveness of this type of reactor, it was necessary to go to higher temperatures to achieve higher thermal efficiencies and higher power densities to reduce capital costs. This entailed increases in cooling gas pressure and changing from Magnox to stainless steel cladding and from uranium metal to uranium dioxide fuel. This in turn led to the need for an increase in the proportion of U^{235} in the fuel. The resulting design, known as the Advanced Gas-Cooled Reactor, or AGR (see Fig 5.11), still uses graphite as the moderator and, as in the later Magnox designs, the steam generators and gas circulators are placed within a combined concrete pressure-vessel/radiation-shield.

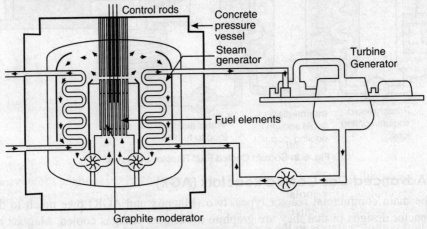

Fig. 5.11: Advanced Gas Cooled Reactor

SUMMARY

This chapter has very briefly brought out the different important components used in the reactors and introduced the reader to the major types of nuclear reactors in operation. In the following chapters more information on the different reactor types is provided.

BIBLIOGRAPHY

1. J.Wood, Nuclear Power, The Institution of Engineering and Technology, London, United Kingdom(2007).
2. www.world-nuclear.org, World Association of Nuclear Reactors.
3. Kenneth Weston, Energy Conversion, 1993, http://www.personal.utulsa.edu/~kenneth-weston/
4. Openshaw Taylor, Nuclear Power Plant, Philosophical Library Inc, New York(1960)[Available from University of Michigan Library]
5. Samuel Glasstone, Alexander Sesonske, Nuclear reactor Engineering, Van Nostrand Reinhold Company; International student edition edition (1969).
6. Kenneth D. Kok (ed.),Handbook of Nuclear Engineering, CRC Press, (2009).
7. Dan Gabriel Cacuci (ed.), Handbook of Nuclear Engineering,Springer (2010).
8. Steven B.,Krivit,Jay H.Lehr, Thomas B. Kingery , Nuclear Energy Encyclopedia: Science, Technology, and Applications, John Wiley (2011).

ASSIGNMENTS

1. Explain the Fermi pile. Collect details of the Fermi pile from literature and present in a seminar.
2. What are the principal components of a nuclear reactor plant.
3. What are the different types of nuclear plants in operation.
4. Compare the fuel, coolant and moderator used in the different types of reactors.

Reactor Thermal Hydraulics

The Nuclear reactor is a source of thermal energy and its heat is used to generate steam to run a steam turbine coupled to a generator to generate electricity. The thermal energy is generated as a result of the fission process in the fissile atoms (Fuel) and the same is transferred to the reactor coolant. This coolant may be boiling water as in a Boiling Water Reactor (BWR) in which case the steam produced in the reactor directly goes to run the Turbine. The coolant may be non boiling water as in the case of Pressurised water Reactor (PWR), in which case the water exchanges heat to water in a steam generator to produce steam. In some reactors the coolant may be a gas like Carbon dioxide or Helium as in Gas Cooled Reactors (GCR), in which case the hot gas would run a gas turbine coupled to a generator. The objective is to remove the heat generated in the fuel without the fuel temperature or coolant temperatures crossing safe limits. Generally, the limit on the Fuel temperature is dictated by the melting point of the fuel material and the limit on coolant temperature is based on the maximum temperature acceptable to the structural materials used in the reactor core. Unlike a fossil fuel plant where the heat generation is nil after stopping of the combustion process, heat continues to be generated in a nuclear reactor after the stopping of the fission reaction due to decay of fission products produced earlier during fission. It is essential to remove this heat known as decay heat to avoid the crossing of temperature limits on clad. This chapter deals with the thermal and hydraulic aspects of the reactor core to maintain the integrity of the core under all normal and abnormal conditions.

6.1 REACTOR HEAT GENERATION

6.1.1 Fission Energy in Reactor

The amount of energy released in fission of a nucleus is related to the net decrease in the mass due to fission. For uranium-235, the average energy released in fission is 200 MeV. This energy is equivalent to three million-times as much energy released with the combustion of carbon of same mass as one uranium-235 atom. The fission energy released appears as kinetic energy of the fission products, kinetic energy of the newborn neutrons, energy of gamma rays emitted during fission and later by fission products, neutrino and beta rays emitted by the fission products following the fission. About 80% of fission energy is in the form of kinetic energy, which manifests as heat and is deposited in the fuel element. The newborn neutrons carry 2.5% of the energy in the form of kinetic energy, and this energy is deposited in the moderator as heat energy. The gamma rays released instantaneously during the fission have about 2.5% energy, and this energy is transferred to fuel and core structures. The delayed energy from beta rays and gamma rays which amounts to 6% of the total is deposited in fuel and structures. All the neutrino associated with beta decay carry about 5% of energy, and this energy escapes the reactor vessel and is non-recoverable, but this loss is compensated by the energy released, about 3.5% within a reactor by excess neutron gamma capture non fission reactions.

In general, the heat generated in the fuel per fission is dependent upon the exact range of particles emitted, depending on the materials used in the reactor and core configuration. The fissile materials uranium-233 and plutonium-239 have slightly different energy release (E) than uranium-235: E (U-233) = 0.98E (U-235) and E (Pu-239) = 1.04E (U-235).

As an example we can calculate the amount of U-235 consumed per day in a thermal reactor operating at power P megawatts. Assuming recoverable energy of 200 MeV (32×10^{-12} J) per fission, the fission rate for a year producing P megawatts power,

Fission rate = P(MW). 10^6 J/MW-s* fissions/($32*10^{-12}$ J)*86400 s/day* 365 days/year

= $9.855 \cdot 10^{23}$ P/year

Burnup rate = $9.855 \cdot 10^{23}$ P/year*235 (g/mole)/$6.22 \cdot 10^{22}$/mole

= 384.58 P g/year

In reality, the fissile material is consumed in fission and in radiative capture. The radiative capture rate is given as:

Radiative capture rate = α x Fission rate,

Where is the ratio of microscopic absorption cross section to the microscopic fission cross section and for U-235 its value is 0.169. Thus:

Consumption rate of U235 = 384.58 (1 + α) P (g) = 449.57P (g)

6.1.2 Heat Generation after shutdown

Up to 50% of the reactor power the heat generation rate can be assumed to follow in proportional to the neutron flux. However, if the reactor is shut down, the reactor core produces heat at ~7% of the reactor power due to delayed neutron fission, fission product decay and activation products from neutron capture. Because of the decay heat, shutdown cooling has to be provided for the reactor core. Immediately after the shut down the reactor power decreases exponential from 7% of the core power with a period of 80 seconds. The heat generation from fission product decay is a result of beta and gamma emission from fission product with decay half-life ranging from microseconds to millions of years.

The decay heat power comes mainly from five sources: (1) unstable fission products, which decay via α, β-, β+ and γ-ray emission to stable isotopes; (2) unstable actinides formed by successive neutron capture reactions in the uranium and plutonium isotopes present in the fuel; (3) fissions induced by delayed neutrons; (4) reactions induced by spontaneous fission neutrons; and (5) structural and cladding materials in the reactor that may have become radioactive. Heat production due to delayed neutron induced fission or spontaneous fission is usually neglected.

6.1.3 Heat Generation in the Moderator

The function of the moderator is to slow down the high energy neutrons liberated in fission, through elastic scattering reactions. This results in heating up the moderator. This heat is about 5% of the heat generated in the reactor. Other effects are due to stopping of beta particles emanating from the fission products and the absorption of gamma rays from various sources.

6.1.4 Heat Generation in Reflectors and Shields

The escaping neutrons from the core get absorbed in the reflector and generate heat. This is besides the gamma radiation which escapes from the core. Neutrons escaping from the reflector deposit all their energies in the shield material as they are slowed down by inelastic and elastic collisions in the shield.

6.1.5 Heat Generation in Structures

Radiation from the core fission process interacts with the core structure and moderator. The most significant energy sources are: fission γ-rays which have high energy, fast neutrons, fission product gamma rays, and capture gamma rays from neutron absorption. The fission gamma rays have very high energy. Although the gamma flux is attenuated as it passes through the metal, the structural elements in the core are quite thin and the attenuation is small.

Radiation absorption into the pressure vessel is an important design problem for PWR. Design pressure of a PWR is high and the reactor vessel is stainless steel-coated carbon steel which loses its ductility in a radiation field. The concern is under accident conditions, large steam line break, or small Loss of Coolant Accident (LOCA), where the hot brittle pressure vessel is repressurized by cold water from the charging pumps. While neutrons can lead to structural damage, it is the gamma rays which contribute to the heating.

6.2 HEAT TRANSPORT IN THE FUEL ELEMENT

There are three modes of heat transfer; conduction, convection and radiation. Conduction requires a media where in adjacent molecules in a solid or fluid collide passing the molecular thermal (translational, rotational, and vibrational) energy from molecular layer to molecular layer. Convection is solely due to the bulk motion of the fluid. The fluid gains thermal energy from a hot surface and is carried to a cold region with fluid flow. The flow can be induced by a pump (forced Convection) or by buoyancy force due to a temperature difference in the fluid (free or natural Convection). Convection can occur in single phase as well as in two-phase flow such as with phase change heat transfer in boiling. The flow can occur external to the surface or wall or internal to a channel or piping. The radiation heat transfer does not require a material medium. Radiation is explained through wave mechanics as electromagnetic waves and in quantum mechanics as particles with speed and a wave frequency. In the fuel pin of a nuclear reactor all these heat transfer mechanisms exist.

In all reactors the fuel is in the form of a pin or plate packed in a metallic cover called Clad. There is a space between the fuel pin and clad to accommodate the fission gases released in fission. If this space is not provided, it is likely that pressure developed due to fission gas release results in failure of clad leading to the contamination of the reactor coolant and increase in radioactivity of the coolant. The heat transfer within the fuel element is by conduction, while the heat transfer in the gap is by free convection and radiation between the fuel outer surface and the inside of clad. Outside the coolant removes heat by forced convection.

6.2.1 General Heat Conduction Equation

The goal of conduction is to determine the temperature field in a medium (such as a fuel rod) and the rate of heat transfer to and from the medium. Typically the media is subjected to non uniform temperature distribution which is a result of a heat source within the medium or heat flux from the boundary of the medium. In this section, various forms of the heat conduction equation that govern the temperature field in a medium and its associated boundary conditions are given.

The general heat conduction can be derived by using the differential analysis for heat diffusion in the Cartesian coordinate system as shown in Figure 6.1. Writing heat balance, the rate of accumulation of thermal energy in an element (of size dx*dy*dz) is equal to the rate of input to the element by conduction, minus the rate of output of thermal energy by conduction, plus the net rate of generation of thermal energy from nuclear reaction.

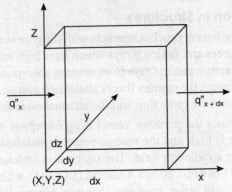

Fig. 6.1: Three dimensional Heat Conduction

The heat diffusion equation:

$$\rho c_p \frac{dT}{dt} = -\frac{dq''_x}{dx} - \frac{dq''_y}{dy} - \frac{dq''_z}{dz} + \dot{q}$$

q'' is the heat flux.
$\dot{q}$ is the volumetric heat generation rate

Using Fourier's law, the heat fluxes (q''_x, q''_y, q''_z) are related to the temperature gradients in the respective directions as

$$q''_x = -k\frac{\partial T}{\partial x}, \quad q''_y = -k\frac{\partial T}{\partial y}, \quad q''_x = -k\frac{\partial T}{\partial z}"$$

In general, the heat conduction equation is written as:

$$\frac{1}{\alpha}\frac{\partial T}{\partial t} = \Delta^2 T + \frac{\dot{q}}{k}$$

where $\alpha = k/\rho c_p$ is called "thermal diffusivity." It indicates how fast the heat diffuses in the medium. Table 6.1 gives the laplacian for all coordinate systems.

Table 6.1: Laplacian for Different Coordinates

Laplacian for Cartesian, Spherical and Cylindrical Coordinates

$\Delta^2 T = \frac{\partial^2 T}{\partial x^2} + \frac{\partial^2 T}{\partial y^2} + \frac{\partial^2 T}{\partial z^2}$	Cartesian
$\Delta^2 T = \frac{1}{r}\frac{\partial}{\partial r}\left(r\frac{\partial T}{\partial r}\right) + \frac{1}{r^2}\frac{\partial^2 T}{\partial \phi^2} + \frac{\partial^2 T}{\partial z^2}$	Cylindrical
$\Delta^2 T = \frac{1}{r^2}\frac{\partial}{\partial r}\left(r^2\frac{\partial T}{\partial r}\right) + \frac{1}{r^2 \sin^2\theta}\frac{\partial^2 T}{\partial \phi^2} + \frac{1}{r^2 \sin^2\theta}\frac{\partial}{\partial \theta}\left(\sin\theta \frac{\partial T}{\partial \theta}\right)$	Spherical

6.2.2 Initial and boundary Conditions

In each equation the dependent variable, T, is a function of four independent variables, (x, y, z, t); (r, ϕ, z, t); (r, ϕ, θ, t) and is a second order, partial differential equation. In order to solve the differential conduction equation for a specific problem, two boundary conditions and one initial condition are required.

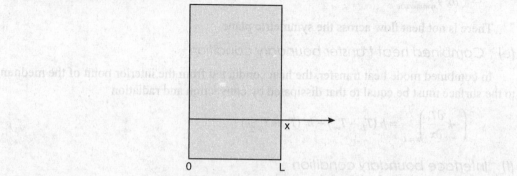

Fig. 6.2: Plane Wall Boundary Condition

6.2.2.1. Boundary Conditions (BC)

Consider a plane wall normal to the x-axis with a thickness L as shown in Figure 6.2. The left-hand surface is located at $x = 0$. The temperature at any point in the wall may be represented by $T(x, y, z, t)$. The common boundary conditions are:

(a) Prescribed surface temperature

At the surface $x = 0$, the temperature is a constant and equals to T_0. The boundary condition is $T(0, y, z, t) = T_0 =$ constant. The prescribed surface temperature is referred to as Drichlet boundary condition.

(b) Prescribed surface heat flux

At the surface $x = 0$, the heat transfer rate per unit area is equal to q'' which is a constant,

$$\left(-k\frac{\partial T}{\partial x}\right)_{x=0} = q''$$

The prescribed heat flux BC is referred as the Neumann boundary condition. If there is no heat flow normal to the surface $x = 0$, the temperature gradient at the surface must be zero.

$$\left(\frac{\partial T}{\partial x}\right)_{x=0} = 0$$

This is adiabatic or insulated surface boundary condition.

(c) Convection boundary condition

The heat conducted outward from the interior point of the wall to the surface $x = L$ must be equal to that dissipated from the surface by convection

$$\left(-k\frac{\partial T}{\partial x}\right)_{x=L} = h(T_L - T_\infty)$$

here h is convection heat transfer coefficient, T_L is the surface temperature at $x = L$ and T_∞ is the ambient temperature of the fluid.

(d) Symmetric boundary condition

If the temperature profile is symmetric about a certain plane, the temperature distribution at this plane must be a maximum or a minimum point and its gradient must be equal to zero.

$$\left(\frac{\partial T}{\partial x}\right)_{symmertric} = 0$$

There is not heat flow across the symmetric plane.

(e) Combined heat transfer boundary condition

In combined mode heat transfer, the heat conducted from the interior point of the medium to the surface must be equal to that dissipated by convection and radiation

$$\left(-k\frac{\partial T}{\partial x}\right)_{x=L} = h(T_L - T_\infty) + h_l(T_L - T_{surf})$$

(f) Interface boundary condition

If two solid plates 1 and 2 are brought together in perfect contact, the heat flux conducted from plate 1 to the interface must be equal to that conducted to plate 2 from the interface and temperatures at the interface of the two plates must be equal

$$-\left(k\frac{dT}{dx}\right)_{1i} = -\left(k\frac{dT}{dx}\right)_{2i}$$

$$T_{1i} = T_{2i}$$

6.2.2.2 Initial Condition

Initial condition is required to find a solution for problems with transient conduction. Initial condition is used to specify the temperature field at initial time. It may specify the initial temperature field at a particular point, surface or region. If the time zero distribution is given as $T_0(x)$, the solution must satisfy $T(x, t = 0) = T_0(x)$.

6.3 HEAT CONDUCTION IN CYLINDRICAL FUEL ROD

Consider one-dimensional steady state heat conduction in a cylinder with internal heat generation and convection boundary condition.

The heat conduction equation in cylindrical coordinates is simplified to:

$$\frac{1}{r}\frac{\partial}{\partial r}\left(kr\cdot\frac{\partial T}{\partial r}\right) + \dot{q} = 0$$

At center of the cylinder $r = 0$, $dT/dr = 0$ (maximum temperature condition) and at surface $r = r_0$, the convection boundary condition

$$\dot{q}(\pi r_0^2 L) = h(2\pi r_0 L)(T_0 - T_a)$$

$$T = T_0 = T_a + \frac{\dot{q}r_0}{2h}$$

In general, the thermal conductivity of the substance is a function of temperature.

Integrating the heat conduction equation twice,

$$\int_{T_s}^{T} k(T')dT' = -\int_{r_0}^{r} \frac{1}{r'} \int_{0}^{r'} r' \dot{q}(r')dr'dr$$

Defining linear heat rate q', which is the rate of heat generation per unit length of fuel rod, $q' = \pi r_0^2 q$, equation, we have the heat conduction

$$\int_{T_s}^{T} k(T')dT' \frac{q'}{4\pi}\left(1 - \frac{r^2}{r_0^2}\right)$$

By assuming the thermal conductivity is independent of temperature and for uniform heat generation rate q', the fuel rod temperature profile is:

$$T - T_0 = \frac{\dot{q}(r_0^2 - r^2)}{4k} = \frac{q'(r_0^2 - r^2)}{4\pi r_0^2 k}.$$

The maximum temperature across the fuel rod occurs at center, $r = 0$

$$T_{max} - T_0 = \frac{qr_0^2}{4k} = \frac{q'}{4\pi k}$$

In terms of the temperature-dependent thermal conductivity, the fuel rod temperature is:

$$\int_{T_s}^{T_{max}} k(T')dT' = \frac{q'}{4\pi}$$

6.4 THERMAL PROPERTIES

The thermal properties of fuel and cladding materials are given in Table 6.2. There is a considerable body of data and correlations in the literature covering the thermal properties of UO2.

Table 6.2: Thermal properties of Fuel and Clad

Property	U	UO2	UC	UN	Zircolloy 2	SS316
Average Density, Kg/m3	$19.04 \cdot 10^3$	$9.67 \cdot 10^3$	$12.97 \cdot 10^3$	$13.06 \cdot 10^3$	$6.5 \cdot 10^3$	$7.8 \cdot 10^3$
Melting Point °C	1330	2800	2390	2800	1850	1400
Thermal Conductivity W/m °C	3.2	3.6	24	21	13(400°C)	23(400°C)
Sp. Heat J/kg °C	116(100°C)	247(100°C)	146(100°C)	–	330(400°C)	580(400°C)
Linear Expansion Coeff. / °C	–	$10.1 \cdot 10^{-4}$	$11.1 \cdot 10^{-4}$	$9.4 \cdot 10^{-4}$	$5.9 \cdot 10^{-4}$	$18 \cdot 10^{-4}$
Tensile Strenth MPa	344-1380	110	62	–		

Source: Adopted from Todreas.N.E, Kazimi.M.S.Nuclear systems I:Thermal Hydraulic Fundamentals, Hemisphere

Uranium oxide fuel is the most widely used fuel in light water and heavy water reactors. The discussions in the following paragraphs pertain to oxide fuel.

Thermal Conductivity: It is the most important thermal property of fuel, which decides the maximum power (W/m) that can be extracted with minimum temperature difference. It decreases as the temperature increases(Fig 6.3).

Porosity Effects: Oxide fuel is generally fabricated by sintering pressed powdered oxide at higher temperatures. By controlling the sintering temperature, fuel of any density up to 90% of the theoretical density can be produced. Conductivity of fuel decreases with increase in voids (pores).Low porosity is therefore essential from considerations of maximum thermal conductivity. However, voids need to be present in fresh fuel to allow space for fission gases as otherwise, fuel may swell and deform or crack the fuel.

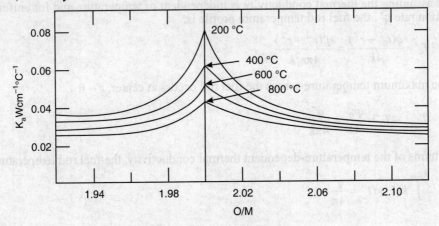

Fig 6.3: Thermal conductivity UO_2

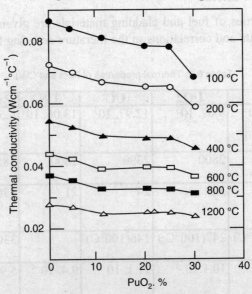

Fig. 6.4: K as function of Pu in $U-PuO_2$

O/M Ratio: Oxygen to metal Ratio of the uranium and plutonium oxides can vary from the theoretical and stoichiometric ratio of 2. This variation affects all the physical properties of the

Reactor thermal Hydraulics

fuel. Departure from the stoichiometric condition happens during the burnup of the fuel. Both hyper and hypostoichiometry reduces K.

Effect of Pu Content: The thermal conductivity of mixed oxide fuel decreases as the content of Pu oxide increases (Fig. 6.4).

Burn Up: Irradiation of fuel induces several changes in the porosity composition and stoichiometry. These changes are small in LWRs; and more in Fast Reactors where Burnup is high.

Effect of Pellet Cracking: Fuel pellet cracking and its relocation in the pellet, cladding gap affects fuel thermal conductivity and gap conductance between fuel and clad.

Fission Gas Release: Accumulated fission gas within clad leads to its pressurization.

Melting Point: The melting point of oxide fuel is ~2840 deg. C. It is affected by the O/M ratio as shown in Fig. 6.5. Impact of Pu to the U oxide on the melting point is given in Fig 6.6. It can be seen that there is a decrease of the melting point with increased Pu content.

Thermal conductivity of 95% Theoretical Density(TD) fresh fuel, 95% TD at 2% burnup and 100% TD fresh fuel are shown in Figure 6.7.

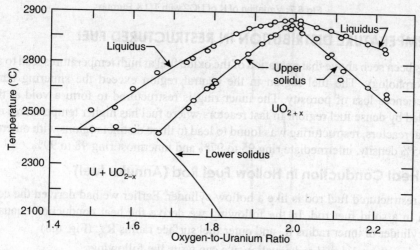

Fig 6.5: Partial Phase Diagram of Uranium Oxide

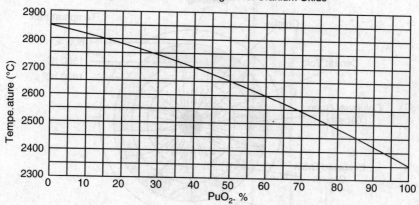

Fig 6.6: Melting Point of Mixed U-PuO_2 As function of PuO_2%

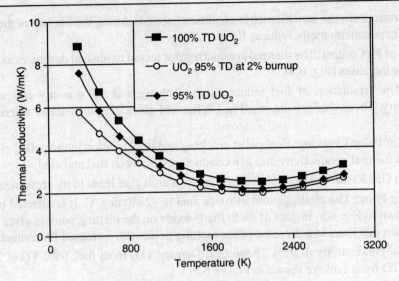

Fig 6.7: Variation of K of UO2 with TD & Burnup

6.5 TEMPERATURE DISTRIBUTION IN RESTRUCTURED FUEL

It has been seen above that operation of the oxide fuel at high temperatures lead to alteration of its morphology. The fuel region in the central region exceed the sintering temperature and experiences loss of porosity. The inner ring is restructured to form a void at the centre surrounded by dense fuel region. In fast reactors where fuel has higher temperatures compared to thermal reactors, restructuring was found to lead to three distinct regions with outermost ring having 95% density, intermediate ring 95 to 97% and innermost ring 98 to 99%.

6.5.1 Heat Conduction in Hollow Fuel Rod (Annular fuel)

The restructured fuel rod is like a hollow cylinder. Earlier we had derived the conduction equation in a solid fuel rod. In the following we derive the heat conduction eqaution of a hollow cylinder of inner radius R_v and outer fuel surface radius R_{fo}. (Fig. 6.8)

Based on the solid fuel rod equation, we can write the following:

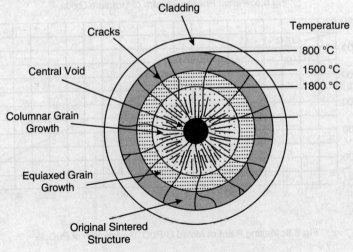

Fig 6.8: Restructured fuel pin of fast reactor

Reactor thermal Hydraulics

$$-\int_{T_{max}}^{T} kdT = \frac{\dot{q}}{4}[r^2 - R_v^2] - \frac{\dot{q}R_v^2}{2} \ell n (r/R_v) \quad (1)$$

This can be rewritten as

$$\int_{T}^{T_{max}} kdT = \frac{\dot{q}r^2}{4}\left\{\left[1 - \left(\frac{R_v}{r}\right)^2\right] - \left(\frac{R_v}{r}\right)^2 \ell n\left(\frac{r}{R_v}\right)^2\right\} \quad (2)$$

The above equation can be used to provide a relation between T_{max}, T_{fo}, R_v and R_{fo}, when the condition $T = T_{fo}$ at $r = R_{fo}$ in applied:

$$\int_{T_{fo}}^{T_{max}} kdT = \frac{\dot{q}r^2}{4}\left\{\left[1 - \left(\frac{R_v}{R_{fo}}\right)^2\right] - \left(\frac{R_v}{R_{fo}}\right)^2 \ell n\left(\frac{R_{fo}}{R_v}\right)^2\right\} \quad (3)$$

The linear heat rate q' is given by

$$q' = \pi (R_{fo}^2 - R_v^2)\dot{q} \quad (4)$$

So that:

$$\dot{q}R_{fo}^2 = \frac{q'}{\pi\left[1 - \left(\frac{R_v}{R_{fo}}\right)^2\right]} \quad (5)$$

Substituting (5) in (3) we get,

$$\int_{T_{fo}}^{T_{max}} kdT = \frac{q'}{4\pi} \times \left[1 - \frac{\ell n(R_{fo}/R_v)^2}{\left(\frac{R_{fo}}{R_v}\right)^2 - 1}\right] \quad (6)$$

If an void factor F_v can be defined based on the ratio of R_{fo} to R_v (α) and void fraction(β)

$$F_v(\alpha, \beta) = 1 - \frac{\ell n(\alpha^2)}{\beta^2(\alpha^2 - 1)} \quad (7)$$

For fuel elements of uniform power density $\beta = 1$. If power density for an inner regions is higher than that of the outer regions $\beta > 1$. This situations is encountered in restructured fuel pellets.

Equation (6) can be rewritten as

$$\int_{T_{fo}}^{T_{max}} kdT = \frac{q'}{4\pi}\left[F_v\left(\frac{R_{fo}}{R_v}, 1\right)\right] \quad (8)$$

Figure 6.9 gives a curve for F_v in terms of α and β.

For the same T_{max}, $q'_{annular} F_v = q'_{solid}$

Provided T_{fo} and K are the same in solid and annular fuel element.

Hence $q'_{annular} > q'_{solid}$. In other words a higher linear power rating is possible with restructured annular fuel.

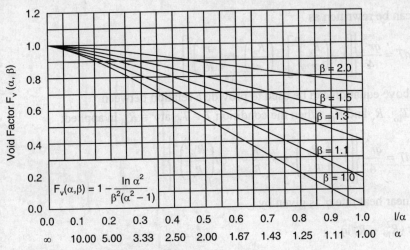

Fig 6.9: Void Factor Function $F_v = f(\alpha, \beta)$

6.6 GAP CONDUCTANCE

The fuel pin is enclosed in a metal clad tube, which acts as the first barrier to the fuel particles outside movement. There is a gap between the outer diameter of the fuel pin and clad which is essential to allow space for fission gases that are released in fission. Otherwise the pressure developed would be high and clad thickness would need to be higher. This would result in more absorption of neutrons in clad leaving fewer neutrons for fission. This gap induces another heat transfer resistance between the fuel pin and clad. In most cases the gap is filled with a conducting gas like helium. Sodium bonded fuels have been used in EBR II a fast reactor built in 1950s in ANL, USA. The gap conductance needs to be accounted for in the heat transfer from fuel to clad.

6.7 HEAT TRANSPORT TO THE COOLANT—SINGLE PHASE

Heat removal from the reactor core is one of the important issues in reactor design and operation. The main objective of heat transfer analysis for single-phase flow in the reactor is to determine the temperature field in a coolant channel such that the reactor operating temperatures are within the specified limits, including the rate of heat transfer to and from a surface or object. Because reactor power densities are typically orders of magnitude higher than other conventional heat sources, the heat removal rate from any given reactor core coolant channel is quite large. The heat transfer to the coolant in single phase thus requires a coolant with a large heat capacity. The coolants should ideally have the following properties:

- Low absorption cross section
- Abundant and inexpensive
- Low melting point
- High boiling point
- Non-corrosive
- High moderating ratio (for thermal reactors)

Reactor thermal Hydraulics

- Radiation and thermal stability
- Low induced radioactivity
- No reaction with turbine working fluid
- High heat transport and transfer coefficient
- Low pumping power

6.7.1 Fluid Flow Characteristics of Coolant

Removal of heat from the fuel involves the flow of fluids around the fuel elements. From the properties of these fluids and their flow characteristics, it is possible to make predictions regarding the heat transfer coefficients for simple geometries. For complex geometry as in nuclear reactors, experimental approach is only feasible. In addition to flow the pressure drops in the various parts of the cooling system are necessary to arrive at the pump specification (flow, head).

When a fluid flows through a pipe at low velocities, the fluid particles move in straight lines parallel to the axis of the pipe, referred to as Laminar flow At the pipe wall due to friction the velocities are minimal, while it is maximum at centre. It is a Parabolic Velocity profile. It obeys Newton's law given by

$$F = \mu A \frac{du}{dy},$$

Where F is the shearing force (Fluid Friction) over an area A, between two parallel fluid layers, moving with a velocity gradient perpendicular to the flow direction and μ is the dynamic viscosity of the fluid. Laminar flow is differentiated from turbulent flow based on Reynolds number Re,

$$Re = \rho V D / \mu$$

Where D is the pipe diameter, v is the mean velocity of the fluid, and p the density. Experiments with different fluids have shown that laminar flow exists as long as the Re is less than 2100. When Re number crosses 4000, the fluid motion becomes turbulent with presence of numerous eddies which cause radial motion of the fluid. Under these conditions Newton's law is not valid. There are three flow regions, one close to the wall where flow is governed by wall friction and hence laminar. This is followed by a transition or buffer layer where some turbulence exists and then fully turbulent flow around the axis. Due to the radial mixing, there is intense momentum transfer and the velocity profile is nearly flat. The ratio of the mean to maximum velocity is around 0.75 at Re of 5000 and increases to 0.81 for Re of 100,000. In the range between Re of 2100 to 4000, there is generally a transition region from purely laminar to turbulent conditions.

6.7.2 Heat Transfer to Coolant

The heat transfer from coolant to any solid surface is given through Newton's cooling law in terms of heat flux:

$$q'' = h(T_s - T_b)$$

where Ts is the surface temperature, T_b is the bulk temperature of the coolant and h (W/m2 °K) is the heat transfer coefficient. The heat transfer coefficient is dependent on coolant properties, flow parameters such as velocity field, temperature field in solid surface and coolant, solid surface condition and geometry.

The heat transfer coefficient h is defined through non-dimensional Nusselt number as:

$Nu = h.D_h/K = f(Re, Pr, Gr, \mu_w, \mu_b)$

where D_h is hydraulics diameter, μ is viscosity, and subscripts w and b refer to wall and bulk, respectively.

The hydraulic diameter D_h is referred to as equivalent diameter when we deal with heat transfer. For circular geometries like pipes, D_h is same as D_{pipe}. For non circular geometries it is given by

$De = 4*$ Cross section of stream/wetted perimeter of duct

The non-dimensional numbers Re, Pr, and Gr are defined as:

$Re = \rho VD/\mu$, Reynolds number,

$Pr = \rho c_p/k$, Prandtl number,

$Gr = g\beta\rho^2(Tw - Tb)^3/\mu^2$, Grashof number,

where ρ is the density, and β is temperature coefficient of expansion.

In most commercial reactors, water is used as the coolant. Water also serves as a moderator in these reactors. Coolants that have been used, in test or commercial applications, include:

- Light water (H_2O) or heavy water (D_2^0)
- Liquid metal sodium, potassium, or NaK (an alloy combination of sodium and potassium))
- Liquid organics (e.g., ethanol, propane, pentane, benzene heptane)
- Air, helium, or carbon dioxide property data of common coolant fluids is given in table 6.3

TABLE 6.3: Property Data For Various Coolants*

Property	Coolant				
	Na	NaK	Hg	Pb	H_2O
T_{melt} (°C)	98	18	−38	328	0
T_{boil} (°C)	880	826	357	1743	100
c_p (kJ/kg.°C)	1.3	1.2	0.14	0.14	4.2
k (W/m.°C)	75	26	12	14	0.7
h (W/m².°C)	36000	20000	32000	23000	17000

*For 3.3m/s velocity in a 25mm duct.

Based on wide experimentation a correlation for heat transfer coefficient of water, named Dittus Boelters equation has been widely used

$Nu = 0.023\ Re^{0.8}\ Pr^{0.43}$

The above equation is applicable for Re>10000 and 0.7<Pr>120. The Pr range includes essentially all liquids and gases except liquid metals, which have Pr-0.001. The liquid metals possess high thermal conductivity, low viscosity and high heat capacity. In liquid metals molecular conduction itself provides 70% of the heat transfer unlike '0.2% to heat transfer in water. The essential difference in heat transfer between liquid metals and ordinary fluids is explained by the fig.6.10.

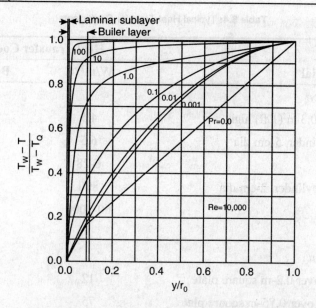

Fig 6.10: Boundary layer Temperature distribution as f(Pr)

Here the approach of the fluid temperature to the tube wall temperature is represented by the dimensionless quantity, $(T_w-T)/(T_w-T_o)$, where T_w and T_o are the wall and centerline temperatures respectively. The abscissa y/r_0, where y is the distance from the wall at which fluid temperature is T and r_0 is the tube radius. The curves are given for different Pr number fluids at a Re of 10000.

For $Pr = 1$, the resistance to heat transfer occurs in the laminar sublayer and in the buffer layer and there is little variation in the temperature further as we move towards centre. For Liquid metals ($Pr < 1$), molecular conduction prevails throughout and there is a temperature gradient as we move from the wall to centerline of the heated pipe. Application of momentum heat transfer analogy led to the following correlation for liquid metals.

$Nu = 7 + 0.025\, Pe^{0.8}$

Where $Pe = Re*Pr$

This equation indicates that there are two terms, the first one referring to conduction and the second one pertaining to convection. It may be seen that the contribution of second term is small. In other words increase of velocity of a liquid metal would not result in a higher heat transfer coefficient as with fluids like water. For design purposes, the conservative Lubarsky's correlation has been widely used for liquid metals.

$Nu = 0.625\, Pe^{0.4}$

It leads to lower heat transfer coefficients for 0.005<Pr<0.05 and Re>10000.

Typical values of the heat transfer coefficient for various fluids and processes are shown in Table 6.4.

Table 6.4: Typical Heat Transfer Coefficients

Mode and Material	Heat Transfer Coefficient (h)	
	W/m²°C	Btu/hr ft² °F
Natural Convection		
Air-vertical plate 0.3 m (1 ft) high	4.5	0.75
Air-horizontal cylinder, 5-cm dia	6.5	1.14
Low-pressure gas	6-28	1-5
Water-horizontal cylinder, 2-cm dia	890	157
Liquids	60-1000	10-175
Forced Convection		
Airflow at 2 m/s over 0.2-m square plate	12	2.1
Airflow at 35 m/s over 0.75-m square plate	75	13.2
Air at 2 atm flowing in 2.5-cm dia tube at 10 m/s	65	11.4
Low-pressure gas flow in pipe	6-600	1-100
Water at 0.5 kg/s flowing in 2.5-cm dia tube	3500	616
Water flow in pipes	250-12,000	44-2000
Sodium flow in pipes	2500-25,000	440-4400
Boiling Water		
In a pool or container	2500-35,000	440-6200
Flowing in a tube	5000-100,000	880-17,600
Condensation of Water, 1 atm		
Vertical surfaces	4000-11,300	700-2000
Outside horizontal tubes	9500-25,000	1700-4400

Sources: Holman, J.P. 1990, *Heat Transfer*, p. 13. 7th ed. McGraw-Hill; adapted from Todreas, N.E and M.S. Kazimi, 1981. Nuclear system I thermal hydraulic fundamentals, Hemisphere, New York.

6.8 HEAT TRANSFER COEFFICIENTS IN FREE CONVECTION

In nuclear reactor systems free or natural convection becomes important in removal of decay heat from the core in the event of loss of pumping power. In free convection, the heat is transferred from the heated solid surface to the adjacent fluid layers. The density of the fluid reduces and the layers move up due to buoyancy and create a flow. The prediction of heat transfer coefficient is more complex in free convection conditions. For laminar free convection from either a flat vertical plate or a cylinder, whose surfaces are at uniform temperature and are placed in an infinite medium, the correlation for heat transfer coefficient is:

$Nu_x = hx/k = 0.55(Gr_x.Pr)^{0.25}$

In the turbulent range of free convection, the following correlation is applicable;
$Nu = 0.13(Gr.Pr)^{0.33}$

6.9 HEAT TRANSFER IN BOILING

In Boiling Water Reactors (BWR), the coolant boils and produces a two phase mixture comprising water and steam. The mixture is sent through moisture separator from where steam is sent to the turbine and condensed steam at turbine outlet recirculated back to the reactor. Thus knowledge of boiling heat transfer is essential to the understanding of the reactor operation. In addition generation of steam in the reactor core leads to changes in the absorption and scattering pattern of neutrons, which result in the change of reactivity (Void Coefficient of reactivity) leading to power changes.

To appreciate the boiling heat transfer, it is convenient to consider a heated surface at temperature T_s in contact with a fluid at a temperature of T_m. If the temperature difference between the surface and fluid is increased, the variation in the heat flux q/A can be plotted as shown in fig 6.11. The curve can be divided into different zones, each with a different heat transfer mechanism. Upto the condition when the heated surface temperature remains below its saturation temperature (Zone I), heat transfer is by single phase convection. In zone II, vapor bubbles forms at the heated surface as the surface temperature exceed the saturation temperature. This referred to as nucleate boiling, where formation of bubbles occurs on nuclei, such as solid particles or gas adsorbed on the heated surface or dissolved in the liquid. The steep rise in the slope of the heat flux curve in zone II is the result of motion of vapor bubbles producing

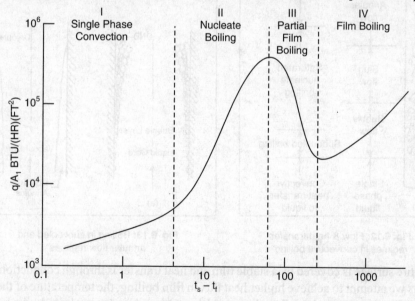

Fig 6.11: Variation of heat Flux with surface-liquid temperature Difference

a good mixing of the liquid. A maximum heat flux is reached when the bubble population increases and bubbles coalesce and form a film, impeding heat transfer. Thus the heat flux decreases, in spite of increase in temperatures. In this zone (III), the film slowly spreads and breaks down. When the film breaks down nucleate boiling takes over in those portions. This is an unstable phase termed partial film boiling. For sufficiently high values of T_s-T_m as in Zone

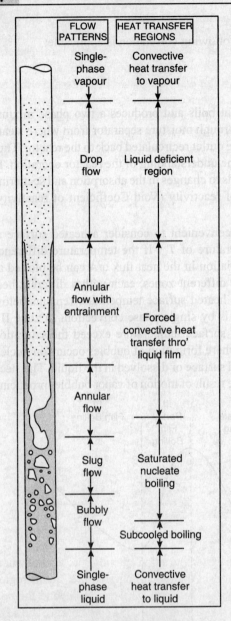

Fig. 6.12: Flow & heat transfer regimes in convective boiling

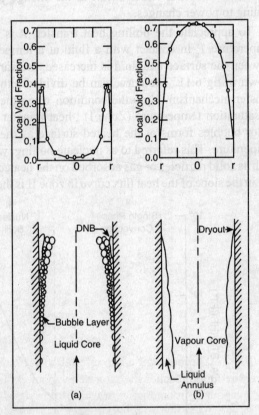

Fig. 6.13: Boiling in subcooled and annular flow regimes

IV, the entire surface is covered by a stable film and heat transfer is through conduction through the film. As we attempt to achieve higher heat flux in film boiling, the temperature of the surface would rise. The maximum heat flux that could be achieved under nucleate boiling is called as critical heat flux (CHF). It is also referred to as burnout heat flux as many experimenters working with electrical heaters observed the burning out of the heater as we approach the maximum heat flux. It is also referred to as departure from nucleate boiling or dryout.

The heat transfer mechanism is governed by the fluid flow pattern. The flow patterns and heat transfer regimes are indicated in Fig 6.12. Water enters the reactor channel as a subcooled liquid (Below Saturation Temperature at the local pressure) and undergoes flow pattern changes causing variations in the heat transfer process. In the single phase liquid region the water temperature at the fuel clad wall is higher than that of the fluid. As the fluid flows picking up heat the inside wall temperature exceeds the saturation temperature under the local pressure conditions, steam bubbles form. This marks the beginning of subcooled boiling. The steam bubbles formed disappear as they move away from the wall, as the temperature there has not reached saturation in the bulk fluid. From the point where temperature across the entire cross section reaches the saturation temperature at the prevailing pressure, bubble formation occurs over the entire cross section. This marks the beginning of nucleate boiling or two phase region.

At higher vapour fractions the flow pattern in the tubes is such that the vapour core exists surrounded by an annulus of water. Heat transfer to water is through the liquid film. If the velocity of vapour in the core is high, turbulence at the vapour-liquid interface causes reduction in the thickness of the liquid film leading finally to dryout. It is seen that such phenomena occur when the heat fluxes are of the order of tens of W/sq.cm as in boiling water reactors.. At higher heat fluxes encountered in pressurised water reactors, fast generation of bubbles results in a large bubble layer, which impedes the heat transfer between the metal surface and liquid. This is referred to as Departure from Nucleate Boiling (DNB). Figure 6.13 depicts the difference between the two phenomena, viz. nucleate boiling and forced convection through liquid film.

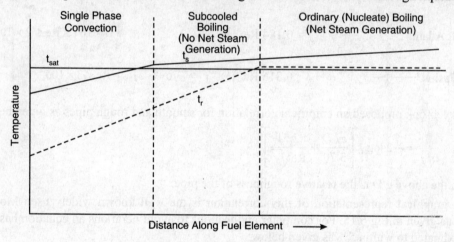

Fig 6.14: Heat Transfer Regimes BWR

The post dryout region also referred to as liquid deficient region extends upto a point where the entire liquid reaches the saturated vapour enthalpy at the prevailing pressure. In this region since vapour is in contact with wall heat transfer rates are small. The region beyond the post dryout regime is the superheated zone where heat transfer is by forced convection. The axial temperatures of fuel clad and water in a BWR is depicted in Fig 6.14.

It must be kept in mind that as we get into the film boiling region, the temperature of the fuel clad will start rising as a consequence of lower heat transfer. It is essential to respect the maximum temperature limits on clad from strength and corrosion considerations as crossing the limits would lead to clad failure and bring the fuel into contact with the coolant. In view of the above there is need to have idea about the critical heat flux under the operating condition in a reactor.

6.10 PRESSURE DROP IN REACTOR SYSTEM

The total pressure drop of a liquid is a sum of the pressure drops due to friction, due to spatial acceleration and gravity or potential drop. In a reactor coolant system, the reactor core comprises many fuel subassemblies each containing a group of fuel elements in the form of plates or cylinders. The coolant flows in-betweens the fuel pins and removes the heat. The gravity head is the product of density p, acceleration due to gravity g and the increase or decrease of level h with respect to a base level. The spatial acceleration or kinetic energy depends on the difference in velocity between inlet and outlet. The friction drop is given by the following equation.

$$\Delta pf = fl\rho v^2/2gD$$

where f is called the friction factor. For Lamina flow f is represented by 64/Re. In the above D is the pipe diameter, l, length, v the mean velocity in the pipe and f the dimensionless friction factor, generally called Darcy's friction factor. Unfortunately another friction factor used in literature is Fanning's friction factor f', where $f = 4f'$.

Expressions for turbulent friction factor f used for different conditions are given below:

$$\frac{1}{\sqrt{f}} = 2\log_{10}(0.398\,\mathrm{Re}\sqrt{f}) \quad \text{(Valid for all Re)}$$

(Karman ± 5% fit) $\quad f = 0.0056 + 0.5\,\mathrm{Re}^{-0.32} \quad 4\times10^3 \le \mathrm{Re} \le 3\times10^6$

McAdams $\quad f = 0.184\,\mathrm{Re}^{-0.2} \quad 3\times10^4 \le \mathrm{Re} \le 2\times10^6$

Blasius $\quad f = 0.316\,\mathrm{Re}^{-0.25} \quad \mathrm{Re} < 3\times10^4$

Colebrook proposed an empirical correlation for smooth and rough pipes as given below:

$$\frac{1}{\sqrt{f}} = -2\log_{10}\left[\frac{e/D}{3.70} + \frac{2.51}{\mathrm{Re}\sqrt{f}}\right]$$

In the above e/D is the relative roughness of the pipe.

A graphical representation of this correlation is the well known widely used Moody's Chart as given in Fig 6.15. For computer applications to avoid iterations an equation has been approximated to within 5% as given below:

$$f = 0.0055(1+[20000*e/D_h+10^6/\mathrm{Re}]**0.33)$$

The coolant velocity will change in areas like header to fuel subassembly (contraction) or in the region at exit of fuel subassembly to top plenum (expansion)(Fig. 6.16). These pressure drops are given by the following equations.

$$\Delta H \text{ (expansion)} = K_e \frac{v_1^2}{2g}$$

$$\Delta H \text{ (contraction)} = K_c \frac{v_2^2}{2g}$$

For abrupt expansion

$$K_e = \left(1 - \frac{D_1^2}{D_2^2}\right)^2,$$

Reactor thermal Hydraulics

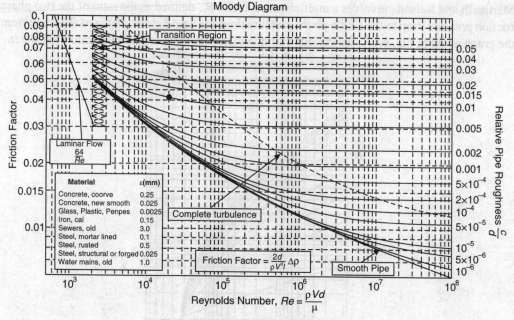

Fig 6.15: Moody's Friction Factor as $f(Re)$

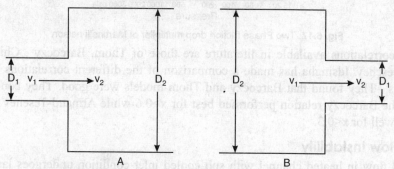

Fig 6.16: Abrupt Expansion and contraction of flow path

Where D_1 and D_2 are the diameters upstream and downstream of the expansion. If D_2 is large as in reactor plenum, Ke is nearly 1.

For abrupt contraction, Kc varies as a function of D_2/D_1 as given below:

D_2/D_1	0.8	0.6	0.4	0.2	0
Kc	0.13	0.28	0.38	0.45	0.50

For the case of entrance into a subassembly from a large header or plenum, Kc approaches 0.5.

Losses in pipe fittings due to change in directions eg. in elbows, curves etc. or due to contraction in valves is represented in terms of the velocity head e.g. $KV^2/2g$. Here values of K can be found from books on fluid flow and Hydraulic Institute standards. It is ~0.25 for a 90° curve and 1.0 for a standard screwed elbow and as high as 10 for fully opened gate valve.

6.11 TWO PHASE PRESSURE DROP

In a boiling channel as in a BWR, it comprises a mixture of water and steam in different proportions across the height of the fuel channel. A methodology based on experiments by

Martinelli and Nelson, involves a multiplication factor R, defined as the ratio of the two phase friction pressure drop to the single phase pressure drop for liquid alone. Value of R is read from the graph given in Fig. 6.17, for various exit qualities with inlet condition as saturated water.

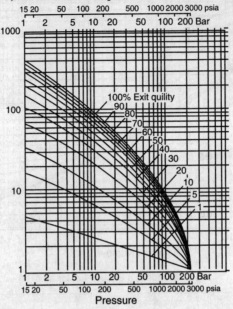

Fig. 6.17: Two Phase friction drop multiplier of Martinelli nelson

Other correlations available in literature are those of Thom, Barcoczy, Chisolm, and Armand-Treschev. Idsingha has made a comparison of the different correlations for ~ 2200 experiments. They found that Barcoczy and Thom models were good. They concluded that for BWR the Barcoczy relation performed best for x>0.6 while Armand-Teschev correlation performed well for x<0.3.

6.11.1 Flow Instability

A fluid flow in heated channel with sub cooled inlet condition undergoes large volume change in a non-uniform manner. As the thermal and hydraulic properties of the flow change continuously along the channel, the flow at any axial point in the channel can never be fully developed thermally or hydrodynamically. Because the flow is not in equilibrium, the flow properties fluctuate upstream and downstream of the point considered often leading to instability. Flow instabilities are of different types depending on the system configuration and operating conditions. In reactors, the oscillations in the flow dramatically reduce the ability of the coolant to remove heat from the core. On the basis of primary features such as oscillation periods, amplitudes, and relationships between pressure drop and flow rate, flow instabilities have been classified into several types. Flow instabilities are caused by self-oscillation of two-phase flow heated channel due to density wave effects. These instabilities are primarily classified as static and dynamic. Some static and dynamic type instabilities occur particularly during startup conditions. In addition to the flow instability BWRs are susceptible to two types of instabilities:

(a) Control system instabilities are due to the malfunction of reactor hardware. Suitable control mechanisms are provided to deal with this type of instability.

(b) Coupled neutronic and thermal hydraulic instabilities (also called "reactivity instabilities") are due to the void and power feedback effects on neutron kinetics and thermal-hydraulics, respectively.

Reactor thermal Hydraulics

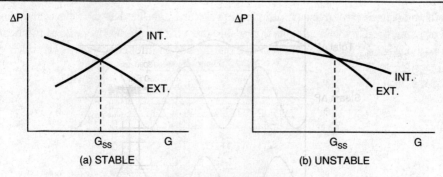

Fig. 6.18: Flow Instability

6.11.1.1 Static Instabilities (Fig 6.18)

A steady flow, rigorously speaking, is one in which the system parameters are functions of the space variables only. Practically, however, they undergo small fluctuations (due to turbulence, nucleation, or slug flow). These fluctuations play a role in triggering several instability phenomena. A flow is stable if, when momentarily disturbed, its new operating conditions tend asymptotically towards the initial ones. A flow is subjected to a static instability if, when the flow conditions change by a small step from the original steady-state ones, another steady state is not possible in the vicinity of the original state. The cause of the phenomenon lies in the steady-state laws and hence, the threshold of the instability can be predicted only by using steady-state laws. A static instability can lead either to a different steady-state condition or to a periodic behavior.

It occurs when the slope of the channel demand pressure-drop - flow-rate curve (internal characteristic of the channel) becomes algebraically smaller than the loop supply pressure-drop - flow-rate curve (external characteristic of the channel). The criterion for this first-order instability is given in Fig. 6.18.

This behavior requires that the channel characteristics exhibit a region where the pressure drop decreases with increasing flow. If the system characteristics have negative slope $\left(\frac{d(\Delta p)}{dG} < 0,\right)$ there is a chance of static instability. Because of this type of instability, multiple steady state operating points and hence flow excursion is possible in water/steam side of steam generators. Modification of system characteristics or the pump characteristics will help to avoid the problems due to static instability. Providing extra pressure drop devices at the steam generator inlet is a common practice to modify the system characteristics.

6.11.1.2 Dynamic Instability

A flow is subject to a dynamic instability when the inertia and other feedback effects have an essential part in the process. The system behaves like a servomechanism and the knowledge of the steady state laws is not sufficient, even for the threshold prediction. The steady-state may be a solution of the equations of the system, but it is not the only solution. There are many types of dynamic instabilities, but the one type that is understood and amenable to analysis is the density wave instability.

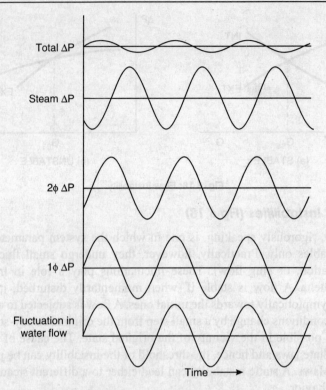

Fig. 6.19: Density wave instability in two-phase flow

A temporary reduction of inlet flow in a heated channel, increases the rate of enthalpy rise, thereby reducing the average density. This disturbance affects the pressure drop as well as heat transfer behavior. For certain combinations of geometrical arrangement, operating conditions and boundary conditions, the perturbations can acquire a 180 deg. out of phase pressure fluctuation at the exit, immediately transmitted to inlet flow rate and become self sustained. For boiling systems oscillations are due to the multiple regenerative feedbacks between flow rate, vapor generation rate and pressure drop. In a simplified way, it is possible to have a total pressure drop across the heated channel, nearly constant with individual pressure drops of single phase, two phase and super heater varying in time due to phase lags arising out of the compressibility variations in different regions (Fig. 6.19). Thus for a given pressure drop across the channels as imposed by the turbine stop valve on the downstream and feed water valve upstream, different channels could be having different flows keeping total water flow same. The periodic oscillation of water flow will induce thermal fatigue due to the change in the critical heat flux region. Typical time period of oscillation for steam/water flow is 5 to 8sec. Large amplitude and temporary reversal of feed water flow has been reported. In some test conditions perturbation in feed water flow and steam flow are in opposite phase in some operating conditions. This explains the phenomenon of self sustained flow oscillations.

6.12 CHOKED FLOW

"Choked flow" (also called "critical flow") is defined in single-phase flow as the flow when the fluid Mach number (which is the ratio between the local fluid velocity and the local sound speed in the fluid) approaches unity. For compressible single-phase flow or for gas-liquid two-phase flow when the Mach number is equal to one, the pressure gradient asymptotically approaches an infinite gradient (shock wave). Similar to the critical flow of compressible single-phase fluid, flow of a two-phase mixture in a channel may also become choked. However, for two-phase flows the sound velocity is frequently smaller than in the flow of either phase separately. The reason is that the mixture has a density of the order of the heavy liquid phase and compressibility close to that of the vapor phase. In reactors during abnormal transients,

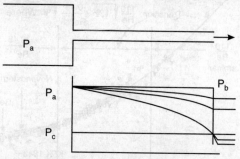

Fig 6.20: Critical/Choked Flow

such as LOCA, choked flow rates limit the flow from reactor vessel. Thus choked flow is important in the analysis of nuclear reactor safety as it determines the coolant inventory loss in case of leakage from primary coolant system in reactor. Consider the flow from a tank containing compressible fluid at P_a discharging under a back pressure P_b (Fig 6.20). As P_b is decreased the flow rate increases until P_b reaches the critical value P_c and then any further decrease in the P_b will not increase the flow rate. This maximum flow rate is said to be choked flow.

6.13 CONDENSATION HEAT TRANSFER

Condensation occurs in many industrial applications, including nuclear reactors. In the latter, condensation occurs in the suppression pool when steam is injected and in turbine condensers. Condensation is classified in the following classes:

(1) Surface filmwise condensation: where the vapor condenses in drops which grow by further condensation and coalesce to form a continuous liquid film and usually flows downward by gravity.

(2) Surface dropwise condensation: if the surface is non-wetting rivulets of condensate flow away and new drops then begin to form resulting in dropwise condensation.

The dominant form of condensation is film condensation, and most of industrial systems employ this form condensation. The local heat transfer coefficients for dropwise condensation are often an order of magnitude greater than those for filmwise condensation. However, it is difficult to maintain the surface to have dropwise condensation. Rates of heat transfer for film condensation can be predicted as a function of bulk and surface temperatures, total bulk pressure, surface and liquid film characteristics, bulk velocity and the presence of non-condensable gases.

6.14 PRESSURE DROP IN ROD BUNDLES

The total pressure drop along a reactor core includes (1) entrance and exit pressure losses between the vessel plena and the core internals, (2) the friction pressure drop along the fuel rods and (3) the form losses due to the presence of spacers. The entrance and exit losses are those (described earlier) due to a sudden change in flow area. Hence attention here is focused on the friction along the rod bundles and the effect of the spacers.

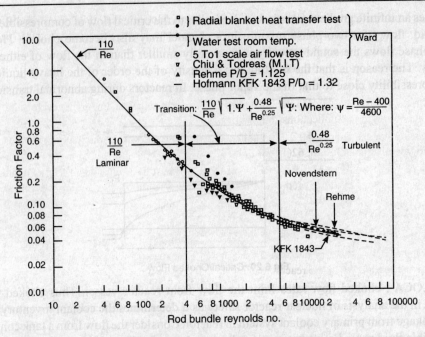

Fig. 6.21: Friction factor for fuel pin bundle

Correlations for friction factor f in rod bundles have been arrived at by many experimenters. An excellent review of all pressure drop correlations for wire wrapped fuel bundle has been recently presented by Bubellis and Schikorr. Fig.6.21 shows some of the experimental results. It can be seen that there is a smooth transition from laminar to turbulent flow for wire wrapped rod bundles, which is in contrast to the more conventional friction factor for round tubes, where a discontinuity is there in the transition region.

During coast down of the pump after power failure, the flow comes down to natural convection conditions. A basic question arises about the use of steady state friction and form loss coefficients for the full range of flow. A.A.Bishop has reviewed data on the friction factors and established that differences in form loss and friction during a transient tend to neutralize each other as they are in opposite directions. Thus, steady state turbulent friction factors can be used for transient conditions.

SUMMARY

This chapter has taken the reader through the applications of heat transfer and fluid flow to reactor design. The various correlations of heat transfer in rod bundles are compiled for sake of completion. They are same as for other applications. It is hoped that with this chapter the reader would be in a better position to follow the contents of further chapters.

BIBLIOGRAPHY

1. Neil Todreas, Mujid S Kazimi, Nuclear Systems Volume I: Thermal Hydraulic Fundamentals, Taylor 86 Francis (1989).
2. J.P. Holman, Heat Transfer, eighth Edition, McGraw Hill(1997).
3. F.P.Incropera and D.P.Dewitt, Fundamentals of Heat and Mass Transfer, John Wiley, Fifth Edition(2005).

4. J.G.Collier , Convective Boiling and Condensation, McGraw Hill(1981).Y.Hsu and R.Graham , Transport Processes in Boiling and Two Phase Systems, Hemisphere Publishing(1976).
5. M.El Wakil , Nuclear Power Engineering ,McGraw Hill, New York.
6. Kenneth D. Kok (ed.) ,Handbook of Nuclear Engineering, CRC Press, (2009).
7. Dan Gabriel Cacuci (ed.) , Handbook of Nuclear Engineering ,Springer (2010).
8. Steven B. ,Krivit , Jay H.Lehr , Thomas B. Kingery , Nuclear Energy Encyclopedia: Science, Technology, and Applications, John Wiley (2011).
9. Bishop.A.A. , Todreas.N. , Hydraulic Chracteristics of wire wrapped rod bundles, Nuclear Engineering and Design, 62(1980) 271-293.
10. Bubellis.E. ,Schkorr ,M. , Review and proposal for best fit of wire wrapped fuel bundle friction factor and pressure drop predictions using various correlations, Nuclear Engineering and Design ,238(2008) 3299-3320.

ASSIGNMENTS

1. What are the sources of heat generation in a nuclear reactor. What is its contribution in terms of the thermal power of the reactor.
2. Derive the heat conduction equation in a cylindrical fuel rod with heat generation.
3. What are the desirable properties of a nuclear fuel. What are the heat transfer properties of fuel that affect its performance.
4. What do you mean by hydraulic diameter.
5. How does the temperature distribution in the boundary layer close to the heat transfer surface differ for water and liquid metals. Give the form of Nusse;lt number correlations for water and sodium, with explanation.
6. What are the important properties that are necessary for a good heat transfer coolant in a nuclear reactor.
7. Explain the different flow and heat transfer regimes in Boiling heat transfer of water.
8. What are the causes of instability in two phase systems. Explain static and dynamic instability.
9. What is the need to know pressure drop of coolant in a reactor system.
10. How does the friction factor in rod bundles differ from that in pipes/tubes.
11. Explain briefly the processes of condensation heat transfer and its relevance in nuclear reactor design.
12. Explain choked flow.

7

Pressurised Water Reactors

7.0 INTRODUCTION

Pressurized water reactors (PWRs) constitute a majority of all nuclear power plants and are one of two types of Light water Reactor. In a PWR the primary coolant (water) is pumped under high pressure to the reactor core, then the heated water transfers thermal energy to a steam generator. In contrast to a boiling water reactor, pressure in the primary coolant loop prevents water from boiling within the reactor. All LWRs use ordinary light water as both coolant and neutron moderator. PWRs were originally designed to serve as nuclear submarine power plants and were used in the original design of the commercial power plant at Shippingport Atomic Power Station. PWRs currently operating in the United States are considered Generation II Reactors. Russia's VVER reactors are similar to U.S. PWRs.

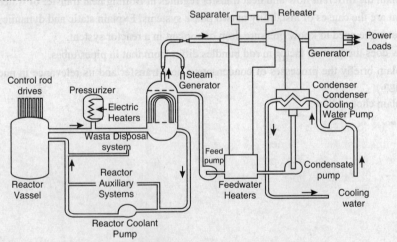

Fig. 7.1: Schematic of PWR

A pressurized water reactor (PWR) is a water-cooled nuclear reactor under sufficient pressure to limit steam generation at the core exit and where the large quantity of heat produced is transferred to a secondary system via a heat exchanger. Several hundred PWRs are used for marine propulsion in aircraft carriers, nuclear submarines and ice breakers. In the US, they were originally designed at the Oak Ridge National Laboratory for use as a nuclear submarine power plant. Follow-on work was conducted by Westinghouse Betits Atomic Power lab. The

Pressurised Water Reactors

first nuclear power plant at Shippingport Atomic Power pressurized water reactor operated pressurized water reactors from 1954 to 1974. In this chapter, we briefly describe the major PWR systems and their role in the operation.

7.1 REACTOR CONFIGURATIONS
7.1.1 Overall System (Fig 7.1)

Nuclear fuel in the core is engaged in a fission chain reaction, which produces heat, heating the water in the primary coolant loop by thermal conduction through the fuel cladding. The primary coolant exiting from the core is cooled in a heat exchanger with a secondary fluid. The overpressure adequate to limit bulk boiling is maintained by an electrically heated pressurizer. In a typical reactor system the primary coolant is circulated through the reactor core by one or more high-pressure pumps.

The hot primary coolant is pumped into a heat exchanger called the steam generator, where heat is transferred across a set of tubes to the lower pressure secondary coolant, which evaporates to pressurized steam. The transfer of heat is accomplished without mixing the two fluids, which is desirable since the primary coolant might become radioactive. Some common steam generator arrangements are u-tubes or single pass heat exchangers.

In a nuclear power station, the pressurized steam is fed through a steam turbine which drives an electrical generator connected to the electric grid for distribution. After passing through the turbine the secondary coolant (water-steam mixture) is cooled down and condensed in a condenser. The condenser converts the steam to a liquid so that it can be pumped back into the steam generator, and maintains a vacuum at the turbine outlet so that the pressure drop across the turbine, and hence the energy extracted from the steam, is maximized. Before being fed into the steam generator, the condensed steam (referred to as feedwater) is preheated in order to minimize thermal shock.

The steam generated has other uses besides power generation. In nuclear ships and submarines, the steam is fed through a steam turbine connected to a set of speed reduction gears to a shaft used for propulsion. Direct mechanical action by expansion of the steam can be used for a steam-powered aircraft catapult or similar applications. District heating by the steam is used in some countries and direct heating is applied to internal plant applications.

Two things are characteristic for the pressurized water reactor (PWR) when compared with other reactor types: coolant loop separation from the steam system and pressure inside the primary coolant loop. In a PWR, there are two separate coolant loops (primary and secondary), which are both filled with demineralized/deionized water. A boiling water reactor, by contrast, has only one coolant loop. The pressure in the primary coolant loop is typically 15-16 MPa (150-160 bar), which is notably higher than in other nuclear reactors, and nearly twice that of a boiling water reactor (BWR). As an effect of this, only localized boiling occurs and steam will recondense promptly in the bulk fluid. By contrast, in a boiling water reactor the primary coolant is designed to boil.

7.1.2 Coolant

Light Water is used as the primary coolant in a PWR. It enters the bottom of the reactor core at about 275 °C (530 °F) and is heated as it flows upwards through the reactor core to a temperature of about 315 °C (600 °F). The water remains liquid despite the high temperature due to the high pressure in the primary coolant loop, usually around 155 b(15.5 MPa 153 atm) . In water, the critical point occurs at around 647 K (374 °C or 705 °F) and 22.064 MPa (218 atm).

Pressure in the primary circuit is maintained by a pressuriser, a separate vessel that is connected to the primary circuit and partially filled with water which is heated to the saturation temperature (boiling point) for the desired pressure by submerged electrical heaters. The pressurizer is also used to act as a surge tank for the system taking up the level variations in the system. To achieve a pressure of 155 b, the pressurizer temperature is maintained at 345 °C, which gives a subcooling margin (the difference between the pressurizer temperature and the highest temperature in the reactor core) of 30 °C. Thermal transients in the reactor coolant system result in large swings in pressurizer liquid volume; total pressurizer volume is designed around absorbing these transients without uncovering the heaters or emptying the pressurizer. Pressure transients in the primary coolant system manifest as temperature transients in the pressurizer and are controlled through the use of automatic heaters and water spray, which raise and lower pressurizer temperature, respectively. Automated pressure control valves (called power operated relief valves) and safety valves, connected to the top of the pressurizer, can open to control and maintain pressure.

To achieve maximum heat transfer, the primary circuit temperature, pressure and flow rate are arranged such that sub-cooled nucleate boiling takes place as the coolant passes over the nuclear fuel rods.

The coolant is pumped around the primary circuit by powerful pumps, which can consume up to 6 MW each. After picking up heat as it passes through the reactor core, the primary coolant transfers heat in a steam generator to water in a lower pressure secondary circuit, evaporating the secondary coolant to saturated steam — in most designs 6.2 MPa (60 atm), 275 °C (530 °F) — for use in the steam turbine. The cooled primary coolant is then returned to the reactor vessel to be heated again.

7.1.3 Moderator

Pressurized water reactors, like thermal reactor designs, require the fast fission neutrons to be slowed down (a process called moderation or thermalization) in order to interact with the nuclear fuel and sustain the chain reaction. In PWRs the coolant water is used as a moderator by letting the neutrons undergo multiple collisions with light hydrogen atoms in the water, losing speed in the process. This "moderating" of neutrons will happen more often when the water is denser (more collisions will occur). The use of water as a moderator is an important safety feature of PWRs, as an increase in temperature may cause the water to turn to steam-thereby reducing the extent to which neutrons are slowed down and hence reducing the reactivity in the reactor. Therefore, if reactivity increases beyond normal, the reduced moderation of neutrons will cause the chain reaction to slow down, producing less heat. This property, known as the negative temperature coefficient of reactivity, makes PWR reactors very stable.

In contrast, the RBMK reactor design used at Chernobyl, which uses graphite instead of water as the moderator and uses boiling water as the coolant, has a large positive thermal coefficient of reactivity that increases heat generation when coolant water temperatures increase. This makes the RBMK design less stable than pressurized water reactors. In addition to its property of slowing down neutrons when serving as a moderator, water also has a property of absorbing neutrons, albeit to a lesser degree. When the coolant water temperature increases, the boiling increases, which creates voids. Thus there is less water to absorb thermal neutrons that have already been slowed down by the graphite moderator, causing an increase in reactivity. This property is called the void coefficient of reactivity, and in an RBMK reactor like Chernobyl, the void coefficient is positive, and fairly large, causing rapid transients. This design characteristic of the RBMK reactor is generally seen as one of several causes of the Chernobyl accident.

PWRs are designed to be maintained in an under moderated state, meaning that there is room for increased water volume or density to further increase moderation, because if moderation were near saturation, then a reduction in density of the moderator/coolant could reduce neutron absorption significantly while reducing moderation only slightly, making the void coefficient positive. Also, light water is actually a somewhat stronger moderator of neutrons than heavy water, though heavy water's neutron absorption is much lower. Because of these two facts, light water reactors have a relatively small moderator volume and therefore have compact cores.

7.1.4 Core Configuration

To date, fuel elements for central station power plants have typically consisted of partly enriched uranium dioxide encapsulated in metal tubes. After enrichment the uranium dioxide (UO_2) powder is fired in a high-temperature, sintering furnace to create hard, ceramic pellets of enriched uranium dioxide. The cylindrical pellets are then clad in a corrosion-resistant zirconium metal alloy (Zircaloy) which is backfilled with helium to aid heat conduction and detect leakages. Zircaloy is chosen because of its mechanical properties and its low absorption cross section. The finished fuel rods are grouped in fuel assemblies, called fuel bundles that are then used to build the core of the reactor. The tubes are assembled into bundles and cooled by water flowing parallel to their axes. A typical PWR has fuel assemblies of 200 to 300 rods each, and a large reactor would have about 150-250 such assemblies with 80-100 tonnes of uranium in all. Generally, the fuel bundles consist of fuel rods bundled 14×14 to 17×17. Present PWR's are rated in the range of 900 to 1,600 MW_e. PWR fuel bundles are about 4 meters in length.

Refuelling for most commercial PWRs is on an 18-24 month cycle. Approximately one third of the core is replaced during each refuelling, though some more modern refuelling schemes may reduce refuel time to a few days and allow refuelling to occur on a shorter periodicity.

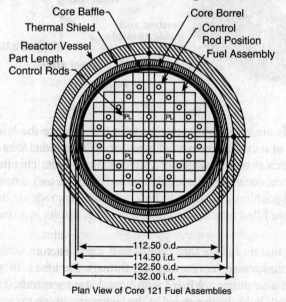

Fig. 7.2: Cross section of PWR Core

Figure 7.2 shows a horizontal cross section of a typical reactor. The fuel assemblies are placed in a configuration approximating a right circular cylinder; the larger the reactor, the more closely a circular cross section is approached. The assemblies are retained in position by a core baffle attached to the core barrel, which is both the structural support for the baffle and the core support plate on which the fuel assemblies rest. The core barrel is surrounded by a thicker ring of metal designated as the thermal shield. The shield and water gaps between the core and vessel thermalize and attenuate the fast neutron flux emanating from the core. In addition, the shield attenuates the gamma flux; this serves to keep thermal stresses, due to gamma and neutron heating, and total neutron exposure of the reactor vessel within acceptable limits.

In American and West European designs, the fuel rods are assembled into large square arrays, which are held in position by spring clips on eggcrate grids spaced about 2 ft apart along the length of the assembly (Fig. 7.3). Fuel tubes are omitted in a number of locations and are replaced by hollow guide tubes that provide the structural support needed to tie the assembly together. The top and bottom nozzles are attached to these guide tubes.

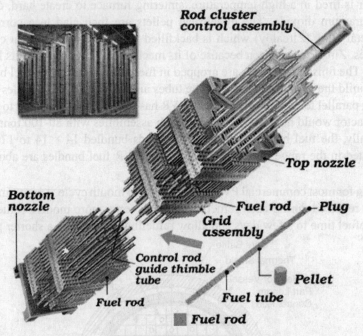

Fig. 7.3: PWR Fuel Assembly

The control rods are clusters of absorber rods that fit in to the hollow guide tube. Each control rod member of a cluster consists of a sealed stainless-steel tube filled with a neutron-absorbing material such as boron carbide or a silver-indium-cadmium alloy. All fuel assemblies are identical and, hence, each has hollow guide tubes, although only a fraction of the assemblies are under control rod positions. In some designs, four adjacent rods are omitted from the lattice and the resulting space filled by a single guide tube. This results in a smaller number of larger control rods.

The guide tubes that do not contain control rods are most commonly filled with plugging clusters to prevent unnecessary coolant bypass through the tubes. In some cases, the guide tubes of a number of assemblies are filled with burnable poison rods. These rods can provide the additional reactivity hold-down needed at the beginning of the cycle when the fuel loading scheme used leads to high discharge burnup.

In earlier designs, cruciform-shaped control rods, fitted into cutouts at the edges of the fuel assemblies, were used (Fig 7.4). To prevent large water gaps when the control rods were

Pressurised Water Reactors

withdrawn, the rods were fitted with non-absorbing extensions. When the control rods were inserted, the extensions projected below the core, necessitating a longer reactor vessel. The individual water slots left by the newer cluster design are small and rod extensions are not needed.

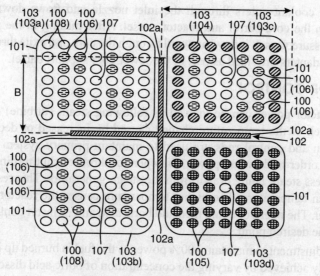

Fig. 7.4: Cruciform Control Rods

7.1.5 Reactor Vessel

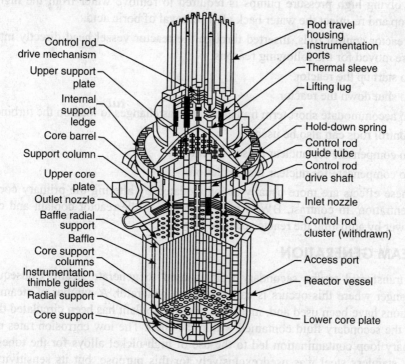

Fig. 7.5: PWR Reactor Vessel

The vertical reactor vessel cross section (Fig. 7.5) illustrates the structural arrangements and coolant flow paths. The core barrel is attached to the lower core support barrel, which is hung from a ledge near the top of the vessel. The upper core support plate and upper control rod shrouds are attached to the upper core support barrel, which fits closely within the lowest support barrel and is supported by it. Both upper and lower core support barrels are designed so they can be removed.

The primary coolant enters through the inlet nozzle and flows downward through the passages between the core barrel and reactor vessel. After turning in the plenum area at the bottom of the pressure vessel, it flows upward through the core and exits through outlet nozzles via close fitting adapters provided as part of the lower core support barrel.

7.2 REACTOR CONTROL

In PWRs reactor power can be viewed as following steam (turbine) demand due to the reactivity feedback of the temperature change caused by increased or decreased steam flow. Boron in solution and control rods are used to maintain primary system temperature at the desired point. In order to decrease power, the operator throttles shut turbine inlet valves. This would result in less steam being drawn from the steam generators. This results in the primary loop increasing in temperature. The higher temperature causes the reactor to fission less and decrease in power. The operator could then add boric acid and/or insert control rods to decrease temperature to the desired point.

Reactivity adjustment to maintain 100% power as the fuel is burned up in most commercial PWRs is normally achieved by varying the concentration of boric acid dissolved in the primary reactor coolant. Boron readily absorbs neutrons and increasing or decreasing its concentration in the reactor coolant will therefore affect the neutron activity correspondingly. An entire control system involving high pressure pumps is required to remove water from the high pressure primary loop and re-inject the water back after removal of boric acid.

The reactor control rods, inserted through the reactor vessel head directly into the fuel bundles, are moved for the following reasons:

- To start up the reactor.
- To shut down the reactor.
- To accommodate short term transients such as changes to load on the turbine.

The control rods can also be used:

- To compensate for nuclear poison inventory.
- To compensate for nuclear fuel depletion.

But these effects are more usually accommodated by altering the primary coolant boric acid concentration. In contrast, BWRs have no boron in the reactor coolant and control the reactor power by adjusting the reactor coolant flow rate.

7.3 STEAM GENERATION

Heat transferred to the secondary fluid is used to generate steam. Consequently, the heat exchanger where this occurs is called a steam generator. A number of steam generator configurations have been used and, in all, the primary coolant has been circulated through the tubes and the secondary fluid contained within the shell. The low corrosion rates required to limit primary loop contamination led to the use of high-nickel alloys for the tubes. Initially, 300-series stainless steel was used exclusively for this purpose, but its sensitivity to stress corrosion led to the use of Inconel in many more recent units. Excessive corrosion rates have also been encountered with Inconel under some conditions. The difficulty appears to be overcome by replacing the phosphate water treatment, originally used for the secondary system, by all-volatile chemistry (ammonia for pH control, hydrazine for oxygen scavenging). However, in some units that operated with phosphate chemistry for an appreciable time, it was found that some corrosion difficulties persisted even after changing to all-volatile chemistry.

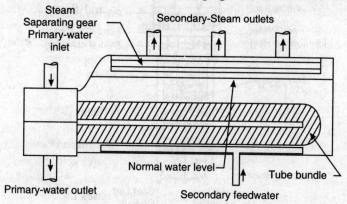

Fig. 7.6: Horizontal Steam Generator-VVER

Horizontal steam generator units are closest to non-nuclear boiler practice. Single-drum or reboiler units (Fig. 7.6), where steam separation takes place in the same drum as the bundle, can be used. Units of this type are still used in the Russian VVER designs. However, capacity is limited by the maximum drum size that can be fabricated or shipped and by the achievable circulation ratio. Higher capacity can be obtained by using multiple drum units such as those shown in Fig 7.7. Here, steam is generated in the lower drum containing the tube bundle and a two-phase mixture flow upward through the risers to the upper drum where the steam is separated. Primary separation takes place at the vortex created in the separators at the entrance to the drum; water is returned to the lower drum via the down-comers. Units of this type were used at Shippingport; the first PWR in commercial service. Although current (1995) U.S. and West European plants have abandoned horizontal steam generators, a Japanese design of simplified PWR system has proposed returning to these units. Such units are stated to have the advantages of freedom from sludge buildup on tube plates and a better capability for withstanding seismic events. Modular vertical steam generators were originally developed to provide compact, high-efficiency units for marine propulsion systems. Nuclear steam generators of this type account for the majority of units installed in large central station plants through 1994. All these units have used a U tube bundle configuration.(Fig 7.8).

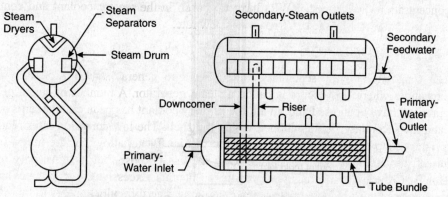

Fig. 7.7: Horizontal multiple drum Steam Generator

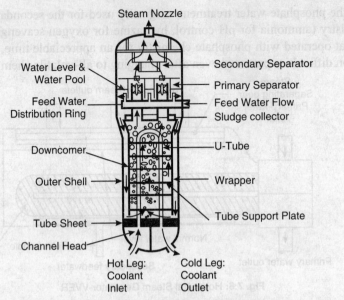

Fig. 7.8: Vertical U tube PWR Steam Generator

Some of the later model steam generators are equipped with an integral pre-heater located on the cold leg of the tube bundle. Such a pre-heater increases the secondary surface temperature in the boiling region and produces higher pressure steam. Feed water is introduced into the pre-heater through a nozzle in the lower shell. The feed water flows through a specially baffled region before it joins the main recirculation flow.

The desire to produce superheated steam has led to the development of the once-through steam generator design. A sectional side view of such a unit is show in Fig. 7.9.

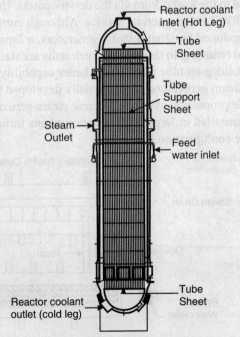

Fig. 7.9: Once Through Steam Generator

The reactor coolant enters the once-through generator at the upper plenum, flows through Inconel tubes, and exits from the lower plenum. The entering feed water is sprayed from an annular ring and mixed with steam bled from the upper end of the boiler section. The mixture flows down the annular down comer and into the tube bank. No distinct water level is present since the mixture quality gradually increases to 100% and superheating starts. The superheater region is baffled to obtain the cross flow and high velocities needed for efficient heat transfer to steam. The superheated steam exits through outlet nozzles near the upper end of the shell.

7.4 OTHER PRIMARY SYSTEM COMPONENTS

In addition to the reactor vessel and steam generators, the primary system contains a pressurizer, large reactor-coolant circulating pumps, and connecting piping. In the most common arrangement, each coolant circulating loop contains a steam generator and pump. In modern designs based on the use of a vertical U-tube steam generator, a medium size plant (e.g., 600 MWe) would contain two loops whereas a large plant (e.g., 1000 MWe) would contain four such loops. The pumps are located on the cold legs (primary fluid outlet lines from the steam generators). In plants using once-through steam generators, the outlet plenum of each steam generator is connected to two circulating pumps. A large PWR of this type would have four circulating pumps, but only two steam generators.

7.4.1 Primary Coolant Pumps

In a reactor system for submarine propulsion, leakage from the primary system cannot be tolerated. Because all mechanical pump seals allow at least some small leakage, canned motor pumps, which have no shaft seals, are used. In such a pump, the entire pump and driving motor are completely enclosed within a pressurized casing. To keep the motor windings dry, the stator and rotor are canned within thin sheets of nonmagnetic material (i.e., 300-series stainless steel). An auxiliary impeller circulates a small amount of water in the narrow space between the stator and rotor liners so as to provide lubrication for the journal and thrust bearing and carry away motor heat. This coolant exits through a tube that is wrapped around the motor casing. External cooling water is circulated over this coil in the space between the coil and an outer jacket. The rotor is directly connected to the pump impeller. A labyrinth type seal limits the interchange between the fluid being pumped and the fluid in the space between the stator and rotor.

Since the first central station PWRs were derived from naval reactor designs, the early PWRs used canned motor pumps. However, leakage requirements in a central station plant are much less stringent than for a naval reactor and mechanical seals with low leakage rates are acceptable. Since large pumps with mechanical seals are both less expensive and more efficient than canned motor pumps, all second and subsequent generation plants in the United States and Western Europe used mechanical seals.

Both first- and second-generation VVER plants used canned rotor pumps. However, the third generation VVER 1000 plants employ shaft-sealed pumps. These shaft-sealed pumps are attached to a flywheel to provide extended flow coast-down in case of a power loss.

In the usual main circulating pump design (Fig.7.10), high-pressure injection water is used to keep the major leakage through the seal inward rather than outward. A typical 90,000 - gpm pump requires of the order of 8 gpm of injection water. About 5 gpm flows downward along the shaft removing heat conducted by the shaft and parts of the thermal barrier. The remaining injection flow passes upward through and around the water lubricated radial bearing of the seal assembly. A typical shaft seal consists of three face seals operating in a series arrangement.

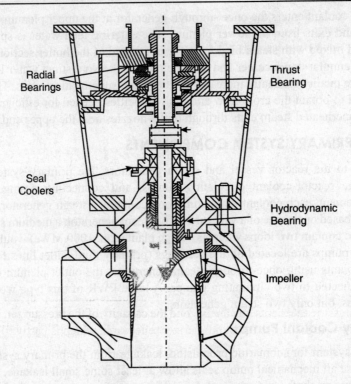

Fig. 7.10: Primary Coolant Pump

Although the leakage rate for the first (primary) seal may be as high as 3 gpm, the leakage rate from the third (outermost) seal is designed to be of the order of 100 cm^3/hr of injection water at operating conditions. Each of the four pumps for a 900-MWe plant would have a capacity of about 90,000 gpm and be operated by a 9000-hp induction motor. The pumps for a 1300-MWe unit would each have a capacity of about 110,000 gpm and be operated by a 12,500-hp motor. The absence of high-pressure injection water leads to pump seal failure. Such a failure produces a small loss-of-coolant accident (LOCA). In view of this possibility, some advanced PWR designs call for the use of canned motor pumps.

7.4.2 Pressurizer

The pressurizer is placed on a pipeline connected to the hot leg of one of the primary circulation loops. A cutaway drawing of a typical unit is shown in Fig 7.11. During steady-state operations, approximately 55 to 60% of the pressurizer volume is occupied by water and 40 to 45% by steam. Electric immersion heaters, located in the lower section of the vessel, keep the water at saturation temperature and maintain a constant system operating pressure.

When the volume of the water in the system increases, the entry of the primary coolant into the pressurizer raises the water level and compresses the steam. The attendant increase in pressure actuates the valves in the spray line. Cold-leg reactor coolant from the outlet side of the pumps then sprays into the steam space and condenses a portion of the steam. In the face of a coolant volume contraction, water from the pressurizer flows outward, thus reducing the pressurizer level and pressure. At the same time, some of the water in the pressurizer flashes to steam. In addition, the immersion heaters come on and further limit the pressure reduction.

Pressurised Water Reactors

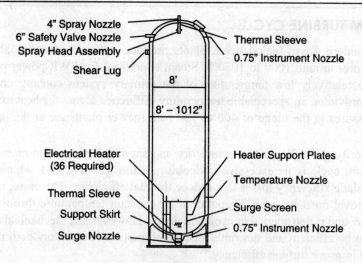

Fig. 7.11: Pressurizer

A power-operated relief valve is connected to a nozzle at the top of the pressurizer to protect against pressure transients that are beyond the capacity of the pressurizer. In such a case, the relief valve discharges steam into a pressurizer relief tank, which is partly filled with water and is where the steam condenses. If the system pressure were to continue to rise, a spring-loaded safety valve would open and thus prevent system failure.

7.5 AUXILIARY SYSTEMS

In addition to the primary loop, several auxiliary systems are necessary for the proper functioning of the plant during normal operation and to provide adequate core cooling during shutdown and accident situations. The major systems and their functions are as follows:

1. *Chemical and volume control system*: Fills and pressurizes system, maintains water level, reduces concentration of contaminants, adjusts concentration of chemical (Boron) poisons added to primary coolant.
2. *Residual heat removal system*: Transfers thermal energy from the reactor coolant during shutdown and refueling operations. System used in conjunction with safety systems during LOCA to remove heat from water being recirculated the by injection system.
3. *Safety injection system*: Rapidly injects water from gas pressurized accumulators during early phase of large LOCA, adds borated water from high head pumps (needed during small LOCA where depressurization is slow), adds large volume of borated water from low head pumps during LOCA (needed during large LOCA with rapid depressurization).
4. *Fuel handling system*: Provides for fuel insertion and removal from the core, provides for fuel storage.
5. *Containment system*: Provides a vessel capable of containing the pressures and temperatures generated by a large LOCA, provides a spray of cold water to limit the containment pressure and to remove iodine from the containment system, provides heat removal from the containment during normal and accident conditions.

7.6 STEAM TURBINE CYCLE

Large, modern, fossil-fueled steam plants produce steam at pressures above 3200 psig and temperatures around 1000 to 1050°F. Steam conditions for PWR power plants are vastly different; the relatively low temperature of the primary system coolant, coupled with the necessity of providing an appreciable temperature difference across the heat exchangers, leads to steam pressures in the range of 400 to 950 psig (newer plants are at the upper end of the range).

The majority of PWR plants produce dry and saturated steam. The main drawback of a saturated steam cycle is its excessive endpoint moisture (20 to 24%), which leads to blade erosion and blade efficiency losses that reduce thermal efficiency. Therefore, some method of moisture removal must be used. Evaporation by constant temperature throttling reduces the steam pressure and is detrimental to cycle efficiency. However, the mechanical steam separator has proved to be efficient and desirable since such moisture separators both reduce endpoint moisture and improve turbine efficiency.

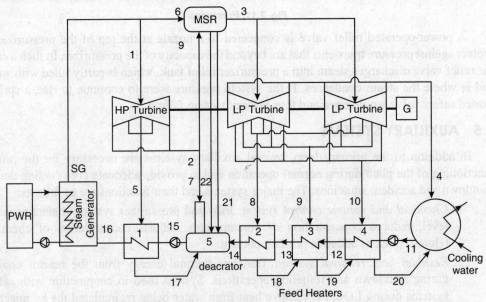

Fig. 7.12: PWR Steam Cycle

Efficiency improvement is enhanced when moisture separation is combined with reheating. Figure 7.12 shows the general arrangement of the elements of a power generation cycle for a medium-size PWR using a combination moisture separator/live-steam re-heater (MSR), which is placed in the crossover piping between the high- and low-pressure turbines. By means of a finned tube heat exchanger, this device removes moisture and transfers heat from steam bypassed ahead of the throttle valve to the main steam. Since superheat increases the inlet temperature of a steam turbine and reduces endpoint moisture, it improves cycle efficiency considerably. In a closed cycle, superheating can be achieved by using either fossil fuel or the once-through steam generator previously described. However, the economics of fossil-fired superheat has not proved attractive and the more modern plants have used saturated steam or superheated steam from a nuclear steam generator. Since the primary heat source is at a relatively low temperature, the amount of superheat that can be supplied by such a steam generator is low; i.e., on the order of 50°F. Thus, the steam cycle followed is quite similar to the dry and saturated cycle

Pressurised Water Reactors

previously described. Moisture separation or moisture separation plus reheat are still required to prevent excessive endpoint moisture.

7.7 CHEMICAL VOLUME CONTROL SYSTEM (FIG. 7.13)

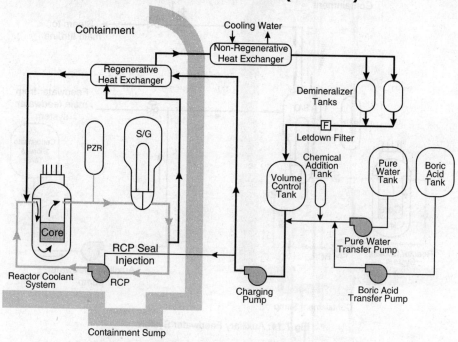

Fig 7.13: CHEMICAL AND VOLUME CONTROL SYSTEM

The chemical and volume control system (CVCS) is a major support system for the reactor coolant system. The functions of the system are to:

- Purify the reactor coolant system using filters and demineralizers,
- Add and remove boron as necessary, and
- Maintain the level of the pressurizer at the desired set point.

A small amount of water is continuously routed through the chemical and volume control system (called letdown). This provides a continuous cleanup of the reactor coolant system which maintains the purity of the coolant and helps to minimize the amount of radioactive material in the coolant. The reactor coolant pump seals prevent the leakage of primary coolant to the containment atmosphere. The chemical and volume control system provides seal injection to keep the seals cool and provide lubrication for the seals.

7.8 RESIDUAL (SHUTDOWN) HEAT REMOVAL CIRCUIT

During normal operation, the heat produced by the fission process is removed by the reactor coolant and transferred to the secondary coolant in the steam generators. Here, the secondary coolant is boiled into steam and sent to the main turbine. Even after the reactor has been shutdown, there is a significant amount of heat produced by the decay of fission products (decay heat). The amount of heat produced by decay heat is sufficient to cause fuel damage if not removed. Therefore, systems must be designed and installed in the plant to remove the decay from the core and transfer that heat to the environment, even in a shutdown plant condition. Also, if it is desired to perform maintenance on reactor coolant system components, the temperature and pressure of the reactor coolant system must be reduced low enough to allow personnel access to the equipment.

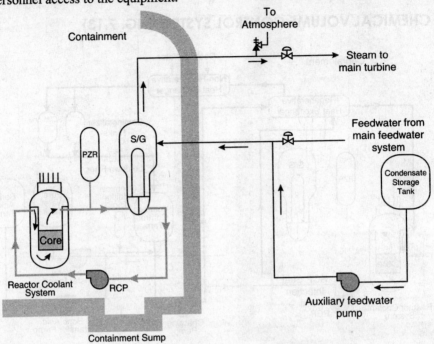

Fig 7.14: Auxiliary Feedwater System

The auxiliary feedwater system and the steam dump system (turbine bypass valves) work together to allow the operators to remove the decay heat from the reactor (Fig. 7.14). The auxiliary feedwater system pumps water from the condensate storage tank to the steam generators. This water is allowed to boil to make steam. The steam can then be dumped to the main condenser through the steam dump valves. The circulating water will then condense the steam and take the heat to the environment. If the steam dump system is not available (for example, no circulating water for the main condenser), the steam can be dumped directly to the atmosphere through the atmospheric relief valves. By using either method of steam removal, the heat is being removed from the reactor coolant system, and the temperature of the reactor coolant system can be reduced to the desired level.

At some point, the decay heat being produced will not be sufficient to generate enough steam in the steam generators to continue the cool down. When the reactor coolant system pressure and temperature have been reduced to within the operational limits, the residual heat removal system (RHR) will be used to continue the cool down by removing heat from the core and transferring it to the (Fig.7.15) environment. This is accomplished by routing some of the reactor coolant through the residual heat removal system heat exchanger, which is cooled by the component cooling water system (CCW). The heat removed by the component cooling water system is then transferred to the service water system in the component cooling water heat exchanger. The heat picked up by the service water system will be transferred directly to the environment from the service water system. The residual heat removal system can be used to cool the plant down to a low enough temperature that personnel can perform any maintenance functions, including refueling.

Pressurised Water Reactors

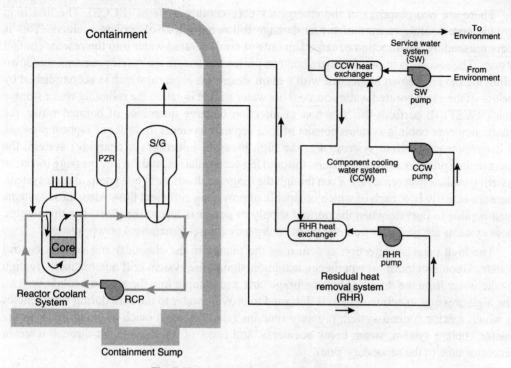

Fig. 7.15: Long term Residual Heat Removal

7.9 EMERGENCY CORE COOLING CIRCUIT

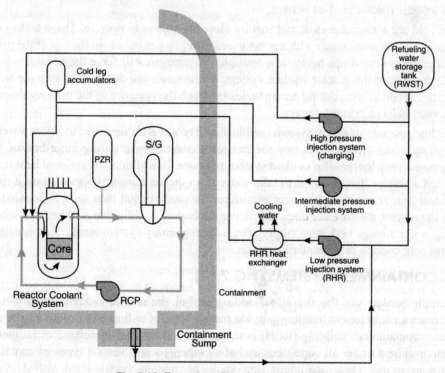

Fig. 7.16: Emergency Core Cooling Systems

There are two purposes of the emergency core cooling systems (ECCS). The first is to provide core cooling to minimize fuel damage following a loss of coolant accident. This is accomplished by the injection of large amounts of cool, borated water into the reactor coolant system. The second is to provide extra neutron poisons to ensure the reactor remains shutdown following the cool down associated with a main steam line rupture, which is accomplished by the use of the same borated water source. This water source is called the refueling water storage tank (RWST). To perform this function of injection of large quantities of borated water, the emergency core cooling systems consist of four separate systems. In order of highest pressure to lowest pressure, these systems are: the high pressure injection (or charging) system, the intermediate pressure injection system, the cold leg accumulators, and the low pressure injection system (residual heat removal). Even though the diagram shows only one pump in each system, there are actually two, each of which is capable of providing sufficient flow. Also, these systems must be able to operate when the normal supply of power is lost to the plant. For this reason, these systems are powered from the plant emergency (diesel generators) power system.

The high pressure injection system uses the pumps in the chemical and volume control system. Upon receipt of an emergency actuation signal, the system will automatically realign to take water from the refueling water storage tank and pump it into the reactor coolant system. The high pressure injection system is designed to provide water to the core during emergencies in which reactor coolant system pressure remains relatively high (such as small break in the reactor coolant system, steam break accidents, and leaks of reactor coolant through a steam generator tube to the secondary side).

The intermediate pressure injection system is also designed for emergencies in which the primary pressure stays relatively high, such as small to intermediate size primary breaks. Upon an emergency start signal, the pumps will take water from the refueling water storage tank and pump it into the reactor coolant system.

The cold leg accumulators do not require electrical power to operate. These tanks contain large amounts of borated water with a pressurized nitrogen gas bubble in the top. If the pressure of the primary system drops below low enough, the nitrogen will force the borated water out of the tank and into the reactor coolant system. These tanks are designed to provide water to the reactor coolant system during emergencies in which the pressure of the primary drops very rapidly, such as large primary breaks.

The low pressure injection system (residual heat removal) is designed to inject water from the refueling water storage tank into the reactor coolant system during large breaks, which would cause a very low reactor coolant system pressure. In addition, the residual heat removal system has a feature that allows it to take water from the containment sump, pump it through the residual heat removal system heat exchanger for cooling, and then send the cooled water back to the reactor for core cooling. This is the method of cooling that will be used when the refueling water storage tank goes empty after a large primary system break. This is called the long term core cooling or recirculation mode.

7.10 CONTAINMENT SYSTEMS (FIG 7.17)

Multiple barriers viz. the zircolloy cladding of fuel, the reactor coolant system boundary and the reactor containment building provide containment of radioactive products. The aim of the reactor containment building (RCB) is to contain and control the release of radioactivity to the environment under all conditions including emergencies. Several types of containment structures are in use. They incorporate steel vaults or concrete vessels lined with steel plates.

Pressurised Water Reactors

Steel vessels are generally cylindrical in shape. Reinforced concrete vessels are cylindrical shaped with hemispherical domes. Steel lining is provided to ensure leak tightness. They include systems to protect the integrity of the containment building by reducing steam pressure (Spray systems) and hydrogen recombiners.

7.10.1 Containment Spray System

Upon the occurrence of either a secondary break or primary break inside the containment building, the containment atmosphere would become filled with steam. To reduce the pressure and temperature of the building, the containment spray system is automatically started. The containment spray pump will take suction from the refueling water storage tank and pump the water into spray rings located in the upper part of the containment. The water droplets, being cooler than the steam, will remove heat from the steam, which will cause the steam to condense. This will cause a reduction in the pressure of the building and will also reduce the temperature of the containment atmosphere (similar to the operation of the pressurizer). Like the residual heat removal system, the containment spray system has the capability to take water from the containment sump if the refueling water storage tank goes empty.

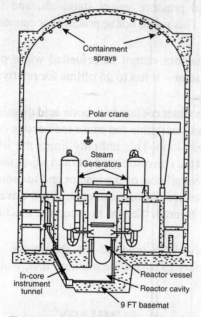

Fig. 7.17: Reactor Containment Building

7.10.2 Hydrogen control in Containment

Under accident conditions zirconium water reactions can result in the release of hydrogen gas to the containment atmosphere. Eventually the gas could accumulate to an explosive hydrogen/oxygen mixture concentration, To control this potential risk, the reactor containment is provided with:

- a hydrogen detection system with alarms to operators
- containment ventilation system to remove pockets of high hydrogen concentration after an accident
- hydrogen recombiners that burn the free hydrogen. The hydrogen/oxygen combination results in water.

7.11 ADVANTAGES OVER BWR

- PWR reactors are very stable due to their tendency to produce less power as temperatures increase; this makes the reactor easier to operate from a stability standpoint.
- PWR turbine cycle loop is separate from the primary loop, so the water in the secondary loop is not contaminated by radioactive materials.
- PWRs can passively scram the reactor in the event that offsite power is lost. The control rods are held by electromagnets and fall by gravity when current is lost.

7.12 PWR TYPICAL ISSUES

- The coolant water must be highly pressurized to remain liquid at high temperatures. This requires high strength piping and a heavy pressure vessel and hence increases construction costs. The higher pressure can increase the consequences of a loss of coolant accident. The reactor pressure vessel is manufactured from ductile steel but, as the plant is operated, neutron flux from the reactor causes this steel to become less ductile. Eventually the ductility of the steel will reach limits determined by the applicable boiler and pressure vessel standards, and the pressure vessel must be repaired or replaced. This might not be practical or economic, and so determine the life of the plant.
- Pressurized water reactors cannot be refuelled while operating. This decreases the availability of the reactor—it has to go offline for relatively long periods of time (~14 days).
- The high temperature water coolant with boric acid dissolved in it is corrosive to carbon steel (but not stainless steel); this can cause radioactive corrosion products to circulate in the primary coolant loop. This not only limits the lifetime of the reactor, but the systems that filter out the corrosion products and adjust the boric acid concentration add significantly to the overall cost of the reactor and to radiation exposure. Occasionally, this has resulted in severe corrosion to control rod drive mechanisms when the boric acid solution leaked through the seal between the mechanisms itself and the primary system.
- Natural uranium is only 0.7% uranium-235, the isotope necessary for thermal reactors. This makes it necessary to enrich the uranium fuel, which increases the costs of fuel production.

SUMMARY

This chapter has brought out the important components of a Pressurised Water Reactor (PWR), which is the largest existing reactor type in operation. The Russian VVER design is similar. The operating experience with this type of reactor has been good.

BIBLIOGRAPHY

1. Pressurized Water Reactor Systems, US Nuclear Regulatory Commission Workshop Manual, USNRC.
2. http://www.nucleartourist.com/type/pwr.htm
3. L.S. Tong, J. Weisman "Thermal Analysis of Pressurized Water Reactors", American Nuclear Society, 3rd Edition, (1996).

4. S. Glasstone and A. Sesonske, Nuclear Reactor Engineering, 3rd Edition, Von Nostrand(1981).
5. Kenneth D. Kok (ed.),Hand book of Nuclear Engineering, CRC Press, (2009).
6. Dan Gabriel Cacuci (ed.), Handbook of Nuclear Engineering,Springer (2010).
7. Steven B.,Krivit, Jay H.Lehr, Thomas B. Kingery, Nuclear Energy Encyclopedia: Science, Technology, and Applications, John Wiley (2011)

ASSIGNMENTS

1. Describe the operation of a PWR with the help of a schematic diagram.
2. Explain the construction of the Fuel bundle of a PWR.
3. What is the role of the pressuriser in a PWR. How does it function.
4. What are the different methods to control power in a PWR.
5. What are the important differences between drum type and once through type steam generators in a PWR.
6. Collect literature on the Three Mile Island Reactor accident and present in a seminar.
7. What are the important auxiliary systems in a PWR.
8. List the issues related to PWR Design.

Boiling Water Reactors

8.0 INTRODUCTION

Boiling Water Reactor (BWR)systems are different from PWRs in that the steam is produced in the core and directly sent to the turbine-generator(Fig 8.1). This is also referred to as a direct cycle. Inside the boiling water reactor (BWR) vessel, a steam water mixture is produced when very pure water (reactor coolant) moves upward through the core absorbing heat. The major difference in the operation of a BWR from other nuclear systems is the steam void formation in the core. The steam-water mixture leaves the top of the core and enters stages of moisture separation, where water droplets are removed before the steam is allowed to enter the steam line. The steam line, in turn, directs the steam to the main turbine causing it to turn the turbine and the attached electrical generator. The unused steam is exhausted to the condenser where it is condensed into water. The resulting water is pumped out of the condenser with a series of pumps and back to the reactor vessel.

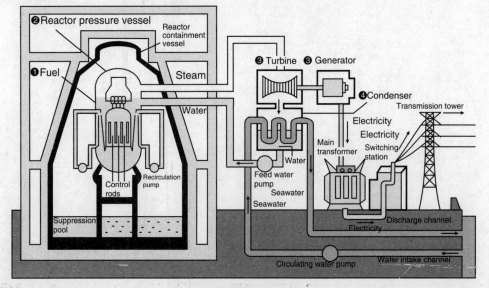

Fig 8.1: BWR Schematic

BWRs offer a cost advantage of eliminating the need for a separate heat exchanger to generate steam. Another feature is that the heat transfer is mostly the latent heat, unlike sensible heat in other power plant cycles. These advantages are countered by the presence of short lived N^{16} in the turbine area as active steam is fed directly to the turbine and this prevents access to the turbine generator till N^{16} activity comes down to acceptable levels. Also the steam pipes would need to shielded for reducing the activity levels in turbine room during normal operation. This chapter gives brief description of the different components of a BWR.

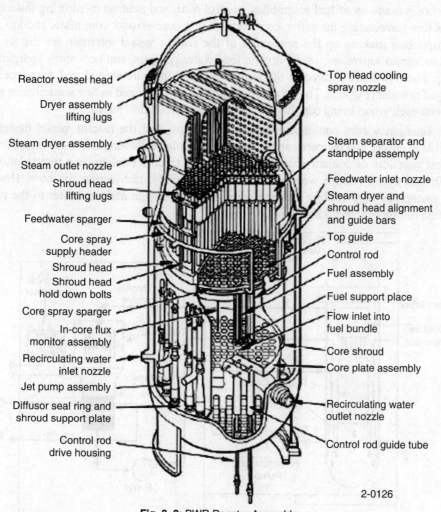

Fig. 8. 2: BWR Reactor Assembly

8.1 BWR REACTOR VESSEL ASSEMBLY

The reactor vessel assembly, shown in fig 8.2, consists of the reactor vessel and its internal components, including the core support structures, core shroud, moisture removal equipment, and jet pump assemblies. The purposes of the reactor vessel assembly are to:
- House the reactor core,
- Serve as part of the reactor coolant pressure boundary,
- Support and align the fuel and control rods,
- Provide a flow path for circulation of coolant past the fuel,
- Remove moisture from the steam exiting the core, and
- Provide a refloodable volume for a loss of coolant accident.

The reactor vessel is vertically mounted and consists of a cylindrical shell with an integral rounded bottom head. The top head is also rounded in shape but is removable via the stud and nut arrangement to facilitate refueling operations. The vessel assembly is supported by the vessel support skirt which is mounted to the reactor vessel support pedestal. The internal components of the reactor vessel are supported from the bottom head and/or vessel wall. The reactor core is made up of fuel assemblies, control rods, and neutron monitoring instruments. The structure surrounding the active core consists of a core shroud, core plate, and top guide. The components making up the remainder of the reactor vessel internals are the jet pump assemblies, steam separators, steam dryers, feed water spargers, and core spray spargers. The jet pump assemblies are located in the region between the core shroud and the vessel wall, submerged in water (Fig 8.3). The jet pump assemblies are arranged in two semicircular groups of ten, with each group being supplied by a separate recirculation pump.

The emergency core cooling systems, penetrations, and the reactor vessel designs are compatible to ensure that the core can be adequately cooled following a loss of reactor coolant. The worst case loss of coolant accident, with respect to core cooling, is a recirculation line break. In this event, reactor water level decreases rapidly, uncovering the core. However, several emergency core cooling systems automatically provide makeup water to the nuclear core within the shroud, providing core cooling.

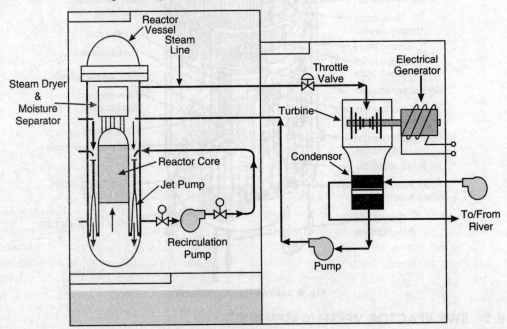

Fig 8.3: BWR Circuit Schematic

8.2 FUEL AND CONTROL ASSEMBLIES

The fuel and control cell assembly (Fig 8.4) is representative for boiling water reactor. Each cell consists of a control rod and four fuel assemblies that surround it. Unlike the pressurized water reactor fuel assemblies, the boiling water reactor fuel bundle is enclosed in a fuel channel to direct coolant up through the fuel assembly and act as a bearing surface for the control rod. In addition, the fuel channel protects the fuel during refueling operations. The power of the core is regulated by movement of bottom entry control rods.

Boiling Water Reactors

8.3 REACTOR WATER CLEANUP SYSTEM (FIG 8.5)

The purpose of the reactor water cleanup system (RWCU) is to maintain a high reactor water quality by removing fission products, corrosion products, and other soluble and insoluble impurities. The reactor water cleanup pump takes water from the recirculation system and the vessel bottom head and pumps the water through heat exchangers to cool the flow. The water is then sent through filter/demineralizers for cleanup. After cleanup, the water is returned to the reactor vessel via the feed water piping.

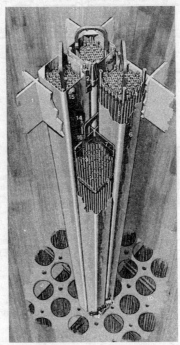

BWR/6 FUEL ASSEMBLIES & CONTROL ROD MODULE
1. Top fuel guide
2. Channel Fastener
3. Upper Tie Plate
4. Expansion Spring
5. Locking Tab
6. Channel
7. Control Rod
8. Fuel Rod
9. Spacer
10. Core Plate Assembly
11. Lower Tie Plate
12. Fuel Support Piece
13. Fuel pelleto
14. End Plug
15. Channel Spacer
16. Plenum spring

GENERAL ELECTRIC

Fig 8.4: Fuel and Control Rods Assembly

8.4 SHUTDOWN/DECAY HEAT REMOVAL (FIG. 8.6)

Heat is removed during normal power operation by generating steam in the reactor vessel and then using that steam to generate electrical energy. When the reactor is shutdown, the core will still continue to generate decay heat. The heat is removed by bypassing the turbine and dumping the steam directly to the condenser. The shutdown cooling mode of the residual heat removal (RHR) system is used to complete the cool down process when pressure decreases to approximately 50 psig. Water is pumped from the reactor recirculation loop through a heat exchanger and back to the reactor via the recirculation loop.

8.5 REACTOR CORE ISOLATION COOLING (FIG. 8.7)

The reactor core isolation cooling (RCIC) system provides makeup water to the reactor vessel for core cooling when the main steam lines are isolated and the normal supply of water to the reactor vessel is lost. The RCIC system consists of a turbine-driven pump, piping, and valves necessary to deliver water to the reactor vessel at operating conditions. The turbine is driven by steam supplied by the main steam lines. The turbine exhaust is routed to the suppression pool. The turbine-driven pump supplies makeup water from the condensate storage tank, with an alternate supply from the suppression pool, to the reactor vessel via the feedwater piping. The system flow rate is approximately equal to the steaming rate 15 minutes after shutdown with design maximum decay heat. Initiation of the system is accomplished automatically on low water level in the reactor vessel or manually by the operator.

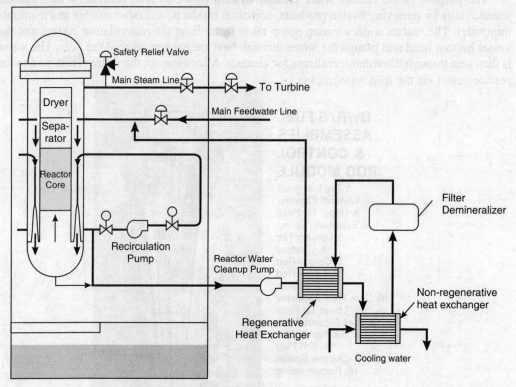

Fig 8.5: Reactor Water Cleanup System

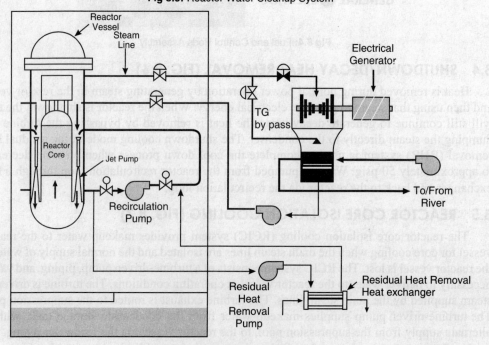

Fig. 8.6: Shutdown Heat Removal

Boiling Water Reactors

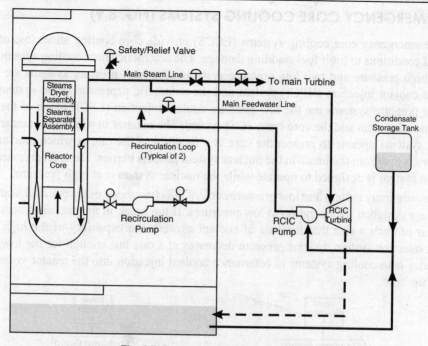

Fig. 8.7: Reactor Core Isolation Cooling

8.6 STANDBY LIQUID CONTROL SYSTEM (FIG 8.8)

The standby liquid control system injects a neutron poison (boron) into the reactor vessel to shutdown the chain reaction, independent of the control rods, and maintains the reactor shutdown as the plant is cooled to maintenance temperatures. The standby liquid control system consists of a heated storage tank, two positive displacement pumps, two explosive valves, and the piping necessary to inject the neutron absorbing solution into the reactor vessel. The standby liquid control system is manually initiated and provides the operator with a relatively slow method of achieving reactor shutdown conditions.

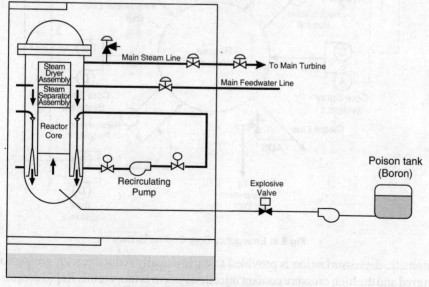

Fig 8.8: Standby Liquid Control System

8.7 EMERGENCY CORE COOLING SYSTEMS (FIG. 8.9)

The emergency core cooling systems (ECCS) provide core cooling under loss of coolant accident conditions to limit fuel cladding damage. The emergency core cooling systems consist of two high pressure and two low pressure systems. The high pressure systems are the high pressure coolant injection (HPCI) system and the automatic depressurization system (ADS). The low pressure systems are the low pressure coolant injection (LPCI) mode of the residual heat removal system and the core spray (CS) system. The manner in which the emergency core cooling systems operate to protect the core is a function of the rate at which reactor coolant inventory is lost from the break in the nuclear system process barrier. The high pressure coolant injection system is designed to operate while the nuclear system is at high pressure.

The core spray system and low pressure coolant injection mode of the residual heat removal system are designed for operation at low pressures. If the break in the nuclear system process barrier is of such a size that the loss of coolant exceeds the capability of the high pressure coolant injection system, reactor pressure decreases at a rate fast enough for the low pressure emergency core cooling systems to commence coolant injection into the reactor vessel in time to cool the core.

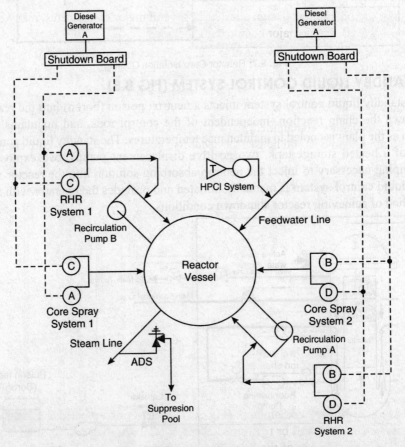

Fig 8.9: Emergency Core Cooling System

Automatic depressurization is provided to automatically reduce reactor pressure if a break has occurred and the high pressure coolant injection system is inoperable. Rapid depressurization of the reactor is desirable to permit flow from the low pressure emergency core cooling systems so that the temperature rise in the core is limited to less than regulatory requirements. If, for a given break size, the high pressure coolant injection system has the capacity to make up for all of the coolant loss, flow from the low pressure emergency core cooling systems is not required for core cooling protection until reactor pressure has decreased below approximately 100 psig. The performance of the emergency core cooling systems as an integrated package can be evaluated by determining what is left after the postulated break and a single failure of one of the emergency core cooling systems.

8.8 BOILING WATER REACTOR CONTAINMENTS

The primary containment package provided for a particular product line is dependent upon the vintage of the plant and the cost-benefit analysis performed prior to the plant being built. During the evolution of the boiling water reactors, three major types of containments were built. All three containment designs use the principle of pressure suppression for loss of coolant accidents. The primary containment is designed to condense steam and to contain fission products released from a loss of coolant accident so that offsite radiation doses are not exceeded and it provides a heat sink and water source.

The Mark I containment design consists of several major components, many of which can be seen on Fig 8.10. These major components include:

- The drywell, which surrounds the reactor vessel and recirculation loops,
- A suppression chamber, which stores a large body of water (suppression pool),
- An interconnecting vent network between the drywell and the suppression chamber, and
- The secondary containment, which surrounds the primary containment (drywell and suppression pool) and houses the spent fuel pool and emergency core cooling systems.

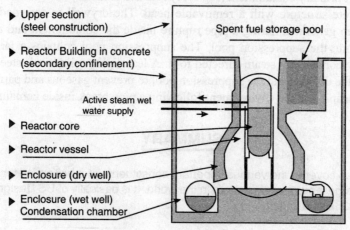

Fig 8.10: Mark I Containment

The Mark II primary containment (Fig 8.11) consists of a steel dome head and either a post-tensioned concrete wall or reinforced concrete wall standing on a base mat of reinforced concrete. The inner surface of the containment is lined with a steel plate that acts as a leak-tight membrane. The containment wall also serves as a support for the floor slabs of the reactor building (secondary containment) and the refueling pools. The drywell, in the form of a frustum of a cone or a truncated cone, is located directly above the suppression pool. The suppression chamber is cylindrical and separated from the drywell by a reinforced concrete slab. The drywell is topped by an elliptical steel dome called a drywell head. The drywell inerted atmosphere is vented into the suppression chamber through as series of downcomer pipes penetrating and supported by the drywell floor.

1 = Primary containment
2 = Drywell
3 = Wetwell
4 = Suppression pool

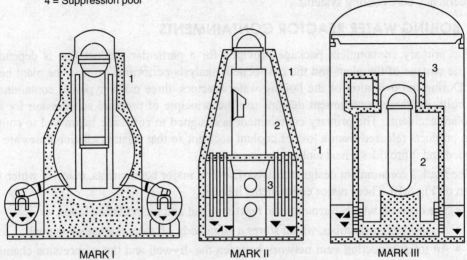

Fig 8.11: Mark I, II, III Containments

The Mark III primary containment (Fig 8.11) consists of the drywell which is a cylindrical, reinforced concrete structure with a removable head. The drywell is designed to withstand and confine steam generated during a pipe rupture inside the containment and to channel the released steam into the suppression pool. The suppression pool contains a large volume of water for rapidly condensing steam directed to it. A leak tight, cylindrical, steel containment vessel surround the drywell and the suppression pool to prevent gaseous and particulate fission products from escaping to the environment following a pipe break inside containment.

SUMMARY

This chapter has covered the various important constituents of a boiling Water reactor. This is the second largest number of reactors in the world. It is basically of US Design. It has been mostly built in USA and Japan.

BIBLIOGRAPHY

1. R.T. Lahey, Jr., F.J. Moody "The Thermal-Hydraulics of Boiling Water Nuclear Reactor", American Nuclear Society, 2 nd Edition(1993).
2. J. G. Collier and J.R. Thome "Convective Boiling and Condensation", Oxford Science Publications, 3 rd Edition, Clarendon Press, Oxford, (1999).
3. N.E. Todreas, M. S. Kazimi "Nuclear Systems II", Taylor & Francis, 1990.

4. E.E. Lewis "Nuclear Power Reactor Safety", John Wiley and Sons, 1977.
5. Kenneth D. Kok (ed.),Handbook of Nuclear Engineering, CRC Press, (2009).
6. Dan Gabriel Cacuci (ed.), Handbook of Nuclear Engineering,Springer (2010).
7. Steven B.,Krivit,Jay H.Lehr, Thomas B. Kingery , Nuclear Energy Encyclopedia: Science, Technology, and Applications, John Wiley (2011)

ASSIGNMENTS

1. Describe the operation of a boiling water reactor with a schematic sketch.
2. What are the advantages and disadvantages of BWR visavis PWR.
3. Explain the role played by the reactor water cleanup system in a BWR.
4. What are the different methods of controlling reactor power in a BWR.
5. Explain the role of emergency cooling systems for a BWR.
6. Collect literature on the Fukushima Reactor accident and present in a seminar.

Pressurised Heavy Water Reactor

9.0 INTRODUCTION

In the 1950s, having proved the feasibility of producing large amounts of energy by nuclear fission in the course of operating research reactors for the production of isotopes, the use of nuclear energy for the commercial production of electricity was under development in a number of countries. Heavy Water Reactor(HWR) programmes were started in Canada, France, Germany, Italy, Japan, Sweden, Switzerland, United Kingdom, United States of America and former USSR. However, it was in Canada that this line of reactors was initially selected as the preferred type, which became known as CANada Deuterium Uranium (CANDU) reactor. The choice was influenced by the early development work in Canada within the Manhattan Project, which took advantage of the superior characteristics of heavy water moderation for the production of plutonium. The attraction for ongoing development was mostly in the comparative simplicity of a system that did not depend on isotopic enrichment of uranium for the fuel. Further simplicity was introduced with the choice of pressure tubes (rather than a pressure vessel as in PWR) to contain the operating pressure. The use of natural uranium and of pressure tubes makes CANDU technology relatively easily accessible.

Each country built research and prototype power reactors, some operating successfully for a number of years, but only the heavy water moderated, heavy water cooled version developed in Canada proceeded to the stage of commercial implementation to become one of the three internationally competitive reactor types available at the end of the 20th century and which has been exported to a number of countries. Light-water-moderated reactors must bear the cost of enriching all of their fuel in U235 throughout their lives. Heavy water moderated reactors avoid this, but must bear the initial cost of producing heavy water. Once produced, however, only minor losses of heavy water occur (typically $\leq 1\%$/a as make-up).

The development of heavy water moderated reactors followed different streams: pressure tube heavy water cooled, pressure vessel heavy water cooled, and pressure tube light water cooled, pressure tube gas cooled and one pressure tube organic cooled design. At the beginning of 2001, 31 heavy water cooled and moderated nuclear power plants were in operation, having a total capacity of 16.5 GW(e), representing roughly 7.76% by number and 4.7% by generating capacity of all current operating reactors. One heavy water moderated, boiling light water cooled reactor was in operation. Six heavy water nuclear power plants were under construction, representing about 18.18% by number and 12.47% by generating capacity of the total units under construction. In total, more than 745 reactor-years of HWR operating experience has been accumulated. A schematic of the reactor is shown in Fig 9.1.

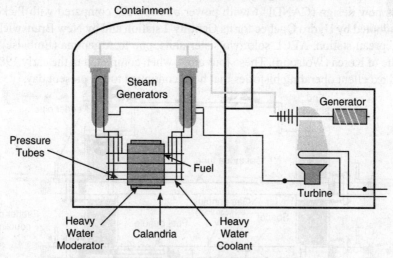

Fig. 9.1: Schematic of CANDU HWR Plant

9.1 GENESIS OF CANDU HWR

The Chalk River site was chosen in 1944 for what was to become the Chalk River Laboratories. At this site, development and construction of the Canadian heavy water moderated research reactors ZEEP (1945), NRX (1947) and NRU (1957), and the development of the laboratories, took place. In 1955, the first small scale prototype heavy water moderated and cooled reactor was committed as a joint undertaking by Atomic Energy of Canada Ltd (AECL), Ontario Hydro (OH (now Ontario Power Generation)) and a private sector company, Canadian General Electric (CGE). The initial design employed a pressure vessel, but in 1957 the design was changed to the pressure tube type. Named the Nuclear Power Demonstration (NPD), this reactor commenced operation in 1962, generating 25 MW of electricity. NPD was followed by the tenfold larger prototype, Douglas Point, which commenced operation in 1967. Located at what later was to become OH's Bruce Nuclear Power Development site on Lake Huron, Douglas Point, together with NPD, established the technological base necessary for the larger commercial CANDU units that followed.

Construction of the first two such commercial units marked the beginning of what currently is OH's eight unit Pickering station. These two units, with a capacity of 500 MW each, were constructed under a tripartite capital financing arrangement between OH, AECL and the Ontario Government. Prior to their completion, OH committed a further two units as a wholly OH investment. The four units came into operation during the period 1971–1973 and established an excellent early performance record. Following the construction of the first four units of Pickering station (Pickering A), OH proceeded with the four unit Bruce A station. Its 800 MW units came into operation in the late 1970s and were followed by four additional units at Pickering (Pickering B) and at Bruce (Bruce B). The latest four unit OH station, Darlington A, started commercial operation in 1991.

Canada made two early entries into the international power reactor supply field. As a first entry, AECL assisted the Indian Department of Atomic Energy (DAE) in the construction of a 200 MW reactor of the Douglas Point type (Rajasthan 1). Following the start of construction of a sister unit (Rajasthan 2), the programme in India was continued by India alone. The second entry was the supply to Pakistan, of a 120 MW CANDU reactor.

AECLs new design (CANDU 6) with power of 600 MW compared with Pickering's 500 MW, was adopted by Hydro Quebec for its Gentilly 2 station and by New Brunswick Power for its Point Lepreau station. AECL sold two sister units, one to Argentina (Embalse) and one to the Republic of Korea (Wolsong). These four units, when completed in the early 1980s, quickly established excellent operating histories that have continued to the present day.

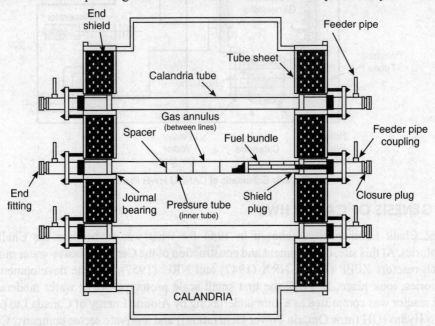

Fig. 9.2: Schematic of Reactor Assembly

9.2 REACTOR

The arrangement of the reactor is shown in Figure 9.2. The cylindrical calandria and end shield assembly is enclosed and supported by the cylindrical shield tank and its end walls. The calandria contains heavy water moderator and reflector; the shield tank contains light water. The heavy water moderator system is independent from the pressurized heavy water heat transport coolant in the fuel channel pressure tubes. The lattice sites, arranged parallel to the horizontal axis, pass through the calandria. Each of the lattice sites accommodates a fuel channel assembly.

Each fuel channel assembly consists of a zirconium-niobium alloy pressure tube, centralized in a calandria tube, and expanded into stainless steel end fittings at both ends. The annulus between the pressure tube and the calandria tube is maintained by annular spacers and is gas-filled to provide thermal insulation. Each of the fuel channel assemblies contains 12 fuel bundles. Fig 9.3 presents an integral view of the calandria and its internals.

9.3 MODERATOR SYSTEMS

The heavy water moderator in the calandria is used to moderate the fast neutrons produced by fission and is circulated through the calandria and moderator heat exchangers (Fig 9.4) to remove the heat generated in the moderator during reactor operation. For a 220 MWe HWR the heat generated in the moderator is 35 MWt. The location of the inlet and outlet nozzles high on the sides of the calandria ensures uniform moderator temperature distribution inside the calandria. The moderator free surface is near the top of the calandria. The operating pressure at the moderator free surface is the normal cover gas system pressure. The moderator system consists of two interconnected circuits, each containing a heat exchanger and a circulation pump. The moderator auxiliary systems include the moderator D_2O collection system, the moderator D_2O sampling system, the moderator liquid poison system, the moderator purification system, and the moderator cover gas system.

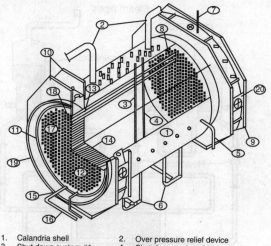

1. Calandria shell
2. Over pressure relief device
3. Shut down system #1
4. Shut down system #1
5. Moderator inlet
6. Moderator outlet
7. Vent pipe
8. Coolant channel assembly
9. End shield
10. End shield support structure ass'y
11. Main shell ass'y
12. Tube sheet F/M side
13. Tube sheet cal side
14. Lattice tube
15. End shield support plate
16. End shield cooling inlet pipes
17. End fitting ass'y
18. Feeder pipes
19. Outer shell
20. Support lug

Fig. 9.3: Integral View of Calandria and Internals

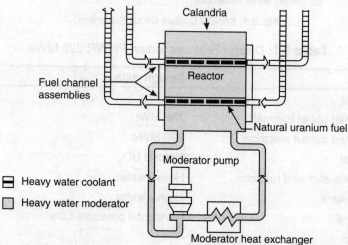

Fig. 9.4: Moderator System Schematic

9.4 HEAT TRANSPORT SYSTEMS

The heat transport system is a single loop with a figure of eight coolant flow pattern. The equipment arrangement with the steam generators and pumps 'in-line' at each end of the reactor results in bi-directional flow through the core (Figure 9.5). The steam generators are of the vertical U-tube type with an integral preheating section. The four heat transport system pumps are vertical single discharge, electric motor driven, centrifugal pumps with multi-stage mechanical shaft seals. The heat transport auxiliary systems include the purification system, the shutdown cooling system, heavy water collection and sampling systems etc.. The design features of the Indian HWR, henceforth referred as the Pressurised Heavy water Reactor (PHWR) of 220 MWe is given in table 9.1.

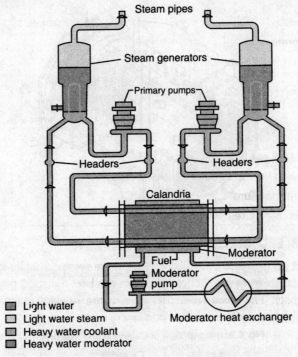

Fig. 9.5: CANDU Coolant Circuit Schematic

Table 9.1: Design Features Indian PHWR 220 MWe

	Design data
A. General	
(i) Rated outlet thermal	756 MWt
(ii) Rated output electrical	220 MWe
(iii) Fuel	Natural UO_2
(iv) Moderator and reflector	Heavy water
(v) Coolant	Heavy water
(vi) Type	Horizontal pressure tube
B. Reactor	
(i) Calandria shell	Horizontal stepped cylinder welded to extensions of end shield
(ii) Calandria shell material	SS-304 L
(iii) End shields	Cylindrical box-type structure integral with calandria shell

Pressurised Heavy Water Reactor

(iv)	End shield material	SS-304 L
(v)	Calandria tubes	
	(a) Quantity	306
	(b) Material	Zircaloy-2
(vi)	Coolant tubes	
	(a) Quantity	306
	(b) Material	Zirconium-2.5% niobium alloy

C. Steam generators

(i)	No. of steam generators	4
(ii)	Type	Vertical U-tubes with integral steam drum (mushroom-shaped)
(iii)	Material	Incoloy-800
(iv)	Steam pressure	3.923 MPa
(v)	No. of tubes per SG	1934
(vi)	Total steam flow to turbine	370 kg/s
(vii)	Steam temperature	250°C
(viii)	Maximum moisture content	0.25%

9.5 FEED AND BLEED CIRCUIT

Inventory control (or pressure control) for the heat transport system is achieved by feed and bleed. Bleed flow is taken from the suction of one of the heat transport pumps and discharged into the bleed condenser. The bleed condenser level control valves maintain the heavy water level in the bleed condenser at the setpoint. One heavy water feed pump is normally operating and takes water from the heavy water storage tank and/or the heat transport purification system and supplies the required flow through feed control valves to the heat transport system via the heat transport pump suction line. The signal to feed or bleed heavy water to or from the heat transport system is based on the pressurizer level during power operation. The heavy water feed pump also provides:
- a cool spray flow to the pressurizer for pressure control,
- a cool spray flow to the bleed condenser for pressure control,
- cool heavy water to the fuelling machine heavy water supply system, and
- a cool flow through the bleed condenser reflux tube bundle for heat recovery.

9.6 FUEL

Fuel characteristics are as follows:
- Use of natural uranium fuel allows the storage and handling of new fuel with minimal criticality concerns because the fuel bundles require heavy water to become critical.
- On-power fuelling means that there is very little reactivity hold-up needed in the reactor control system (and no need for boron in the coolant to hold down reactivity, resulting in a simpler design). The control rod reactivity worth can therefore be kept quite small (2 mk per rod or less).
- The high neutron economy, and hence low reactivity hold-up, of HWRs means that the reactor is very unlikely to become critical after any postulated beyond design-basis severe core damage accident.

- Low remaining fissile content in spent fuel means that there are no criticality concerns in the spent fuel bay.

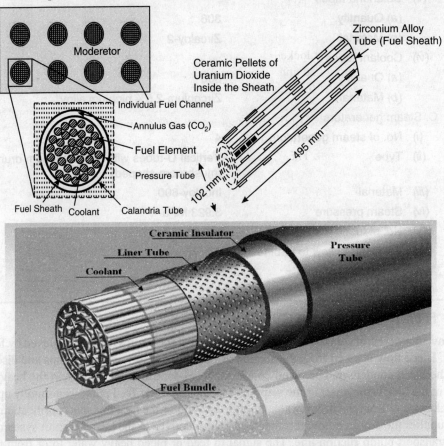

Fig. 9.6: Fuel arrangement in calandria tube and fuel bundle

Each fuel element contains sintered pellets of uranium dioxide with a U235 content of 0.71 wt% in a Zircaloy-4 sheath(Fig 9.6). There is a graphite layer on the inside surface of the sheath. End caps are resistance welded to the ends of the sheaths to seal the element. End plates are resistance welded to the end caps to hold the elements in a bundle assembly. Spacer pads are brazed to the elements at their midpoints, to provide inter-element spacing. Element contact with the pressure tube is prevented by bearing pads brazed near the ends and at the mid-point of each outer element. Beryllium metal is alloyed with the Zircaloy-4 to make the braze joints.

9.7 FUEL HANDLING

On-power refuelling and a failed fuel detection system allow fuel that becomes defective in operation to be located and removed without shutting down the reactor. This reduces the radiation fields from released fission products, allows access to most of the containment while the reactor is operating, and reduces operator doses. As a result of on-power fuelling, the core state does not change after about the first year of operation. Thus, the reactivity characteristics remain constant throughout plant life, resulting in simpler operation and analysis. Ability to couple tools to the fuelling machine allows it to be used for some inspections without necessitating removal of the pressure tube and in some instances without de-fuelling the channels.

Pressurised Heavy Water Reactor

The pressure tube design is convenient for on-power refueling. The fuel inside the individual pressure tubes can be changed using remotely controlled fuelling machines one channel at a time. The CANDU reactors are refueled on-power using two fuelling machines located at opposite ends of the reactor. The fuelling machines are operated from the main control room. For refueling, the fuelling machines are positioned at opposite ends of the fuel channel to be refueled, and locked on to the end fittings to obtain leak-tight joints. After the fuelling machines are aligned and clamped to the channel, the pressures in the machines and channel are equalized. The fuelling machines then remove the channel closures, guide sleeves are installed and injection flows are established from the fuelling machines. The shield plugs are then removed which allows the fuelling machine at the inlet end of the channel to move the fuel string towards the fuelling machine located at the outlet end. A ram adapter is added to the fuelling machine ram. The irradiated fuel bundles are supported with this ram adapter as they are pushed, separated into pairs and stored in the fuelling machine. During refueling, the irradiated fuel bundles are removed from the fuel channel outlet and new fuel bundles added at the inlet. The fuelling machine unloads irradiated fuel bundles through the associated irradiated fuel port which leads to one of the irradiated fuel storage bays located outside the reactor building containment wall.

9.8 REACTOR POWER CONTROL

Total reactor power is controlled automatically by computer from zero power to full power. Liquid zone control compartments, distributed throughout the reactor core in vertical zone control units, provide the primary means to regulate reactivity during normal reactor operation. The zone control units adjust the flux level in any of the reactor zones by adding or removing light water to/from the zone control compartment to provide local control of neutron absorption. The semi-continuous on-power refueling system provides the principal means of long term reactivity control. In addition to the zone control system, adjuster rods, mechanical control absorbers, and the addition of soluble poison to the low temperature moderator are other means available for reactivity control. The adjuster rods adjust the neutron flux for optimum reactor power and fuel burnup. They also compensate for xenon reactivity after a reactor shutdown. The reactor regulating system allows the reactor power to be reduced to about 60 percent of full power and operation continued indefinitely at that level or to be quickly reduced to zero power and then restarted within 35 minutes (which is the xenon override time). Steam discharge to the turbine condenser allows continued reactor operation at reduced power when the turbine or electrical grid connection is not available.

9.9 REACTOR SAFETY

Safety related systems perform the safety functions necessary to maintain the plant in a safe condition during normal operation, and to mitigate events caused by the failure of the normally operating systems or by naturally occurring phenomena (e.g. earthquakes). The safety related systems used for mitigating events include four special safety systems and safety support systems which provide necessary services to the former. The four Special Safety Systems are: Shutdown System Number 1 (SDS1), Shutdown System Number 2 (SDS2), Emergency Core Cooling (ECC), and Containment. Safety related systems are separated into two groups, Group 1 and Group 2, to provide protection against common cause events which impair a number of systems or damage a localized area of the plant (e.g. fires). Group 1 includes most of the systems required for normal operation of the plant as well as two special safety systems. Systems in each group are capable of performing the essential safety functions of reactor shutdown, decay heat removal, and control and monitoring. Each group contains safety support services (e.g. electrical power systems with diesel generators, cooling water systems, and steam generator feedwater systems), which can also provide backup support services to the other group, as required. During normal plant operation, the Group 1 systems generally provide the support services to the systems in both groups.

The systems in each group are designed to be as independent from each other as practicable, to prevent common cause events from affecting systems in both groups. Interconnections between groups are kept to a practical minimum, and provided with suitable isolation devices. Inside the reactor building, components of each group are physically separated by distance or local carriers. Outside the reactor building, the physical interface between groups is designed as a fire barrier, and as a barrier against any flooding that can occur in either of the groups. Fire barriers, flood control, and physical separation of selected components are also provided where necessary within each group.

9.9.1 Shutdown Systems (Fig 9.7)

Two shutdown systems, additional to the regulating system, are provided to shut the reactor down for safety reasons. One system (SDS1) consists of mechanical shutdown rods while the other (SDS2) injects a gadolinium nitrate solution into the moderator via liquid poison injection nozzles. These two safety systems are independent and diverse in concept and are separated both physically and in a control sense to the maximum practical degree from each other and from the reactor regulating system. The two shutdown systems respond automatically to both neutronic and process signals. Either shutdown system, acting on its own, is capable of shutting down the reactor and maintaining it shut down for all design basis events.

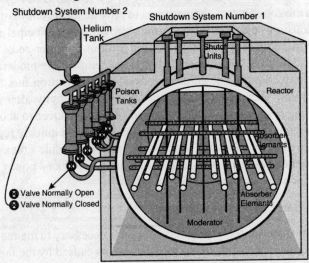

Fig. 9.7: Reactor Shutdown Systems

9.9.2 Shutdown Cooling System

Heat continues to be generated in the reactor after shutdown due to decay of fission products. The residual power is referred as decay power and needs to be removed by the shutdown cooling system. The major functions of the shutdown cooling system are to:

(a) Cool the heat transport system after a reactor shutdown following an initial cool down by steam rejection, to a temperature suitable for maintenance.

(b) Maintain the heat transport system temperature at the maintenance level for any desired length of time.
(c) Provide a means of draining, refilling and level control of the heat transport system to allow maintenance of the heat transport pumps or steam generators.
(d) Cool down the heat transport system from the zero power hot temperature under abnormal conditions.
(e) Provide a long term heat sink after a design basis earthquake, following depletion of feedwater to the steam generators.

9.9.2.1 Emergency Core Cooling System (Fig 9.8)

The purpose of the emergency core cooling system is to replenish the reactor coolant and to assure cooling of the reactor fuel in the unlikely event of a loss-of-coolant accident. Emergency core cooling is provided by injecting light water into the reactor headers. During the initial injection phase, air from the emergency core cooling system gas tanks is utilized to pressurize the emergency core cooling system water tanks and deliver light water to the headers at high pressure. In the long term the emergency core cooling system recirculates water through the reactor and dedicated heat exchangers for decay heat removal. Make-up water is provided from the reserve water tank.

The emergency core cooling system, except for the gas tanks and recovery pumps, is housed within the reactor building.

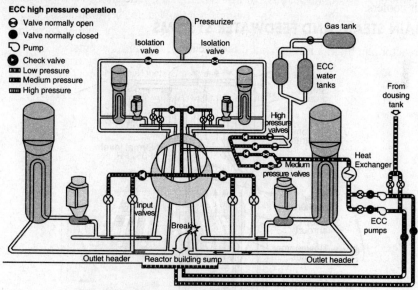

Fig. 9.8: Emergency Core Cooling System

9.10 CONTAINMENT SYSTEM (FIG 9.9)

The containment system consists of the containment envelope and the containment isolaticn. The containment envelope is a pressure-retaining boundary consisting of the reactor building and metal extensions such as airlocks, piping system penetrations and electrical penetration assemblies. The airlock links the containment building with outside and is such that in case of opening only outside air leaks in. The containment envelope is designed to withstand the maximum pressure which could occur following the largest postulated loss-of-coolant accident. Piping systems passing through the envelope are equipped with isolation valves.

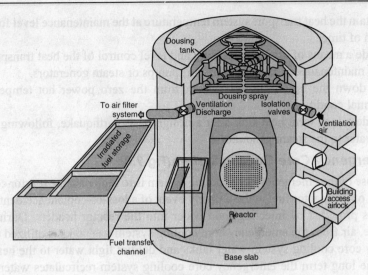

Fig. 9.9: PHWR Containment

The containment isolation system automatically closes all reactor building penetrations open to the containment atmosphere when an increase in containment pressure or radioactivity level is detected. A long-term containment atmosphere heat sink is provided by the reactor building air coolers.

9.11 MAIN STEAM AND FEEDWATER SYSTEMS

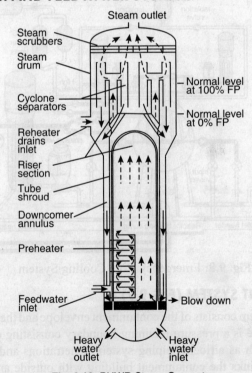

Fig. 9.10: PHWR Steam Generator

Steam generators with integral pre heaters transfer heat from the heavy water reactor coolant of the primary heat transport system side to the light water on the secondary side (Fig 9.10).

Pressurised Heavy Water Reactor

The temperature of the incoming feed water is increased to the boiling point and subsequently evaporated. The steam generators consist of an inverted vertical U-tube bundle installed in a shell. Steam separating equipment is housed in the steam drum at the upper end of the shell. The steam from the boilers is fed by separate steam mains to the turbine steam chest via the turbine stop valves, and its flow is controlled by the governor valves. The steam pressure is normally controlled at a constant value by varying reactor power to match the turbine-generator demand. The condenser steam discharge valves, in combination with the atmospheric steam discharge valves, are sufficient to avoid lifting of the main steam safety valves following a loss of line or a turbine trip and, hence, permit continuation of reactor operation. Main steam safety valves are provided on each steam main to protect the steam system and the steam generators from overpressure.

The feedwater system supplies normal feedwater to the steam generators. The feedwater system comprises the main feedwater pumps on Class IV power and a diesel-driven auxiliary feedwater pump. The feed water is demineralized and preheated light water.

9.11.1 Steam Generator Pressure Control

The steam generator pressure control system enables the reactor power output to track the turbine power output, using the steam generator pressure as the controlled variable. The steam generator pressure controller is a part of the overall plant control system. During normal operation, steam pressure is primarily controlled by adjusting reactor power. If for some reason the reactor regulating system does not allow the reactor to respond to pressure controller demands, or if a reactor power reduction occurs because of a trip, a stepback, or a setback, the reactor set point is controlled directly by the respective reduction signal, and the 'normal' mode of control of steam generator secondary side pressure is interrupted. Steam pressure control switches to the 'alternate' mode of adjusting the plant loads. Similarly, if the operator elects to control the reactor power set point manually, steam pressure control is via plant loads.

When the plant is in the 'normal' mode, the turbine governor valves are controlled through the unit power regulator program; i.e., the unit power regulator calculates what the valve set point should be and pulses to that position. If the plant is in the 'alternate' mode, the steam generator pressure control system controls the turbine in response to the steam pressure error, steam pressure error rate of change, and the rate of change of reactor power.

Atmospheric Steam Discharge Valves are low capacity valves that are used to control steam generator pressure via the steam pressure control program. They are opened in proportion to the pressure error, normally with an offset in the steam pressure set point. These valves may also be used to provide a heat sink during shutdown for decay heat removal when the main condenser is unavailable. Condenser Steam Discharge Valves are capable of discharging up to 70% of full power live steam to the condenser on loss of turbine so that the reactor can continue to operate at the power required to prevent a 'poison-out'. They are also used to discharge steam on a loss of line, or on a turbine trip, so that the main steam safety valves do not lift. During normal operation these valves operate on the pressure control mode. On a turbine trip, they are first opened fully and then returned to the pressure control mode. During reactor shutdown they provide a heat sink through the condenser for decay heat removal.

9.11.2 Steam Generator Level Control

The level in each of the steam generators is controlled individually, as a function of power. Because of safety, range of control and maintenance considerations, each steam generator has a set of three control valves for feedwater control connected in parallel: one small valve to control feedwater during shutdown, startup, and low power operation, and two larger valves to control feedwater for on-power conditions. Each of the two large valves can handle the full power flow requirements. Isolating valves are provided for each control valve. The steam generator level control system balances feedwater to steam flow for all operating conditions. The water level set point is automatically programmed over a set range as a function of load.

SUMMARY

Pressurised Heavy Water Reactors are the third largest group of reactors. They use natural uranium as fuel unlike enriched uranium used in PWR and BWR. Thus this type of reactor has been built in many developing countries like India, where enrichment processes which are power dependant tend to be uneconomical. The different components and their functions have been given in great detail.

BIBLIOGRAPHY

1. https://canteach.candu.org
2. S.S.Bajaj,A.R.Gore, The Indian PHWR, Nuclear Engg. And Design, Vol.236, Issue 8, Pages 701–722,(2006).
3. Indian220 MWe PHWR (IPHWR-220) Status Report for Advanced Nuclear Reactor Designs -Report 74,(2011).
4. http://www.npcil.nic.in/main/KnowledgePortal
5. AERBSAFETY GUIDE NO. AERB/NPP-PHWR/SG/D-10, SAFETY SYSTEMS FOR PRESSURISED HEAVY WATER REACTORS, Atomic Energy Regulatory Board, India,(2005).
6. REVIEW of INDIAN NPPs-POST FUKUSHIMA EVENT, http://www.npcil.nic.in/pdf/presentation.pdf (2011).
7. IAEA-TECDOC-1594,Analysis of Severe Accidents in Pressurized Heavy Water Reactors, IAEA,Vienna,(2008).
8. Kenneth D. Kok (ed.), Handbook of Nuclear Engineering, CRC Press, (2009).
9. Dan Gabriel Cacuci (ed.), Handbook of Nuclear Engineering, Springer (2010).
10. Steven B.,Krivit,Jay H.Lehr, Thomas B. Kingery, Nuclear Energy Encyclopedia: Science, Technology and Applications, John Wiley (2011).
11. S.A. Bharadwaj, The Future 700 MWe pressurised heavy water reactor, Nuclear Engineering and Design, 236(2006)861-871.

ASSIGNMENTS

1. Describe the operation of a PHWR with a schematic sketch.
2. What are the advantages of a PHWR concept over a PWR.
3. Explain the operation of the moderator system in a PHWR. What are the advantages of having separate moderator and coolant systems in a PHWR unlike in a BWR/PWR.
4. How is the pressure maintained in the primary heat transport system in a PHWR.
5. Doesthe PHWR have any advantage with respect to fuel handling as compared to a BWR/PWR.
6. Explain the Shutdown system of a PHWR.
7. Howis shutdown cooling accomplished under different conditions in a PHWR.
8. Whatare the methods by which reactor power control is accomplished in a PHWR.

Gas Cooled Reactor

10.0 INTRODUCTION

A gas-cooled reactor (GCR) uses natural uranium as fuel, graphite as a neutron moderator and carbon dioxide or Helium as coolant. Although there are many other types of reactor cooled by gas, the terms GCR and to a lesser extent gas cooled reactor are particularly used to refer to this type of reactor. Use of natural uranium was attractive for many countries so that they need not pursue the costly enrichment technologies.

There were two main types of GCR:
- The Magnox reactors developed by the United Kingdom.
- The UNGG reactors developed by France.

The main difference between these two types is in the fuel cladding material. Both types were mainly constructed in their countries of origin, with a few export sales: Magnox plants to Italy and Japan, and a UNGG to Spain. Both types used fuel cladding materials that were unsuitable for medium term storage under water, making reprocessing an essential part of the nuclear fuel cycle.

In the UK, the Magnox was replaced by the advanced gas-cooled reactor (AGR), an improved Generation II gas cooled reactors. In France, the UNGG was replaced by the pressurized water reactor(PWR). More recently, GCRs based on the declassified drawings of the early Magnox reactors have been constructed by North Korea at the Yongbyon Nuclear Scientific Research Center. This chapter is devoted to an overview of the gas cooled reactors.

10.1 MAGNOX-GENERAL DESCRIPTION

Magnox reactors are pressurised, carbon dioxide cooled graphite moderated reactors using natural uranium (i.e. unenriched) as fuel and magnox alloy (an alloy—mainly of magnesium with small amounts of aluminum and other metals—used in cladding unenriched uranium metal fuel with a non-oxidising covering to contain fission products) as fuel cladding. Boron -steel control rods were used. The design was continuously refined, and very few units are identical. Early reactors have steel pressure vessels, while later units (Oldbury and Wylfa) are of prestressed concrete; some are cylindrical in design, but most are spherical. Working pressure varies from 6.9 to 19.35 b for the steel pressure vessels, and the two prestressed concrete designs operated at 24.8 and 27 b.

On-load refuelling was considered to be an economically essential part of the design for the civilian Magnox power stations, to maximise power station availability by eliminating refuelling downtime. This was particularly important for Magnox as the unenriched fuel had a low burnup, requiring more frequent changes of fuel than enriched uranium reactors. However the complicated refuelling equipment proved to be less reliable than the reactor systems, and perhaps not advantageous overall.

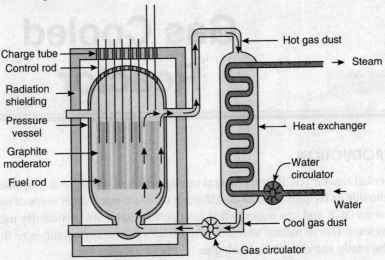

Fig 10.1: Schematic of Magnox Reactor

The first Magnox reactors at Calder Hall were designed principally to produce plutonium for nuclear weapons. The production of plutonium from uranium by irradiation in a pile generates large quantities of heat which must be disposed of, and so generating steam from this heat, which could be used in a turbine to generate electricity, or as process heat in the nearby Windscale works. Once removed from the reactor the used fuel elements are stored in cooling ponds (with the exception of Wylfa which has dry storage in a carbon dioxide atmosphere) where the decay heat is transferred to the pond water, and then removed by the pond water circulation, cooling and filtration system. The fact that fuel elements can only be stored for a limited period in water before the Magnox cladding deteriorates, and must therefore inevitably be reprocessed, added to the costs of the Magnox programme.

The Magnox reactors were considered at the time to have a considerable degree of inherent safety because of their simple design, low power density, and gas coolant. Because of this they were not provided with secondary containment features. A safety design principle at the time was that of the "maximum credible accident", and the assumption was made that if the plant were designed to withstand that, then all other lesser but similar events would be encompassed. Loss of Coolant accidents (at least those considered in the design) would not cause large-scale fuel failure as the Magnox cladding would retain the bulk of the radioactive material, assuming the reactor was rapidly shutdown (a SCRAM), because the decay heat could be removed by natural circulation of air. As the coolant is already a gas, explosive pressure buildup from boiling is not a risk, as happened in the catastrophic steam explosion at the Chernobyl accident. Failure of the reactor shutdown system to rapidly shut down the reactor, or failure of natural circulation, was not considered in the design. In 1967 Chapelcross experienced a fuel melt due to restricted gas flow in an individual channel and, although this was dealt with by the station crew without major incident, this event had not been considered in the design or planned for, and the radioactivity released was greater than anticipated.

Gas Cooled Reactor

In the older steel pressure vessel design, boilers and gas ducting are outside the concrete biological shield. Consequently this design emits a significant amount of direct gamma and neutron radiation, termed direct "shine", from the reactors. For example the most exposed members of the public living near Dungness Magnox reactor in 2002 received 0.56 mSv, over half the International Commission on Radiological Protection(ICRP) recommended maximum radiation dose limit for the public, from direct "shine" alone. The doses from the Oldbury and Wylfa reactors, which have concrete pressure vessels which encapsulate the complete gas circuit, are much lower.

The first Magnox power station, Calder Hall, was the world's first nuclear power station to generate electrical power on an industrial scale. First connection to the grid was on 27 August 1956, and the plant was officially opened by Queen Elizabeth II on 17 October 1956. When the station closed on 31 March 2003, the first reactor had been in use for nearly 47 years.

The first two stations (Calder Hall and Chapelcross) were originally owned by the UKAEA and primarily used in their early life to produce weapons grade plutonium, with two fuel loads per year. From 1964 they were mainly used on commercial fuel cycles and in April 1995 the UK Government announced that all production of plutonium for weapons purposes had ceased.

In operation it was found that there was significant oxidation of mild steel components by the high temperature carbon dioxide coolant, requiring a reduction in operating temperature and power output. For example the Latina reactor was derated in 1969 by 24%, from 210 MWe to 160 MWe, by the reduction of operating temperature from 390 to 360°C.

As of 2011, two Magnox power stations remain in operation, Oldbury and Wylfa. Originally Oldbury was to close in 2008, but in an announcement on 18 December 2008 by the UK Nuclear Decommissioning Authority it was stated that the station would continue to operate for another two years, in order to raise funds to pay for decommissioning. Wylfa has been extended to 2012.

Magnox had the advantage of a low neutron capture cross-section, but had two major disadvantages:

- It limits the maximum temperature, and hence the thermal efficiency, of the plant.
- It reacts with water, preventing long-term storage of spent fuel under water.

Magnox fuel incorporated cooling fins to provide maximum heat transfer despite low operating temperatures, making it expensive to produce. While the use of uranium metal rather than oxide made reprocessing more straightforward and therefore cheaper, the need to reprocess fuel a short time after removal from the reactor meant that the fission product hazard was severe. Expensive remote handling facilities were required to address this danger.

10.2 MAGNOX REACTORS IN OTHER COUNTRIES

Some countries other than UK also built similar designs as given below.

- Three North Korean reactors, all based on the declassified blueprints of the Calder Hall Magnox reactors:
- A small 5 MWe experimental reactor at Yongbyon, operated from 1986 to 1994, and restarted in 2003. Plutonium from this reactor's spent fuel has been used in the North Korea nuclear weapons programme..
- A 50 MWe reactor, also at Yongbyon, whose construction commenced in 1985 but was never finished in accord with the 1994 US-North Korea Agreed framework.
- A 200 MWe reactor at Taechon, construction of which also halted in 1994.

- Nine UNGG power reactors built in France, all now shut down. These were carbon dioxide-cooled, graphite reactors with natural uranium metal fuel, very similar in design and purpose to the British Magnox reactors except that the fuel cladding was magnesium- zirconium alloy.

The accepted term for all of these first-generation, carbon dioxide-cooled, graphite-moderated reactors, including the Magnox and UNGG, is GCR for Gas Cooled Reactor.

The Magnox was replaced in the British power station program by the Advanced Gas Cooled Reactor or AGR, which was derived from it. A key feature of the AGR was the replacement of magnox cladding to allow higher temperatures and greater thermal efficiency. Stainless steel cladding was adopted after many other alloys had been tried and rejected.

10.3 UNGG REACTORS

The UNGG (Uranium Naturel Graphite Gaz) is an obsolete design of nuclear power reactor developed by France. It was graphite moderated, cooled by carbon dioxide, and fueled with natural uranium metal.

It was developed independently of and in parallel to the British Magnox design, and to meet similar requirements. The main difference between the two designs is that UNGG used a horizontal fuel rod orientation, rather than the vertical orientation used in the Magnox reactor. The fuel cladding material was magnesium-zirconium alloy in the UNGG, as opposed to magnesium-aluminum in Magnox. As both claddings react with water, they can be stored in a spent fuel pool for short times only, making short-term reprocessing of the fuel essential, and requiring heavily shielded facilities for this.

The UNGG and the Magnox are the two main types of GCR. A UNGG reactor is often referred to simply as a GCR in English documents, or sometimes loosely as a Magnox. The first generation of French nuclear power stations were UNGGs, as was Vandellos unit 1 in Spain. Of the ten units built, all were shut down by end 1994, most for economic reasons due to staffing costs. The largest was Bugey 1 with a net electrical output of 540 MW. The earlier units, at Chinon and Marcoule, had heat exchangers outside the main pressure vessel; later units (Saint-Laurent, Bugey and Vandellos) moved these heat exchangers to inside the pressure vessel.

10.4 ADVANCED GAS COOLED REACTOR (AGR)

Advanced gas-cooled reactor (AGR) is the second generation of British gas-cooled reactors. The AGR was developed from the Magnox reactor, operating at a higher gas temperature for improved thermal efficiency, requiring stainless steel fuel cladding to withstand the higher temperature. Because the stainless steel fuel cladding has a higher neutron capture cross section than Magnox fuel cans, enriched uranium fuel is needed, with the benefit of higher "burn ups", requiring less frequent refuelling. The first prototype AGR became operational in 1962 but the first commercial AGR did not come on line until 1976.

All AGR power stations are configured with two reactors in a single building. Each reactor has a design thermal power output of 1,500 MWt driving a 660 MWe turbine-alternator set. The various AGR stations produce outputs in the range 555 MWe to 670 MWe though some run at lower than design output due to operational restrictions. The design of the AGR was such that the final steam conditions at the boiler stop valve were identical to that of conventional coal-fired power stations, thus the same design of turbo-generator plant could be used. The mean temperature of the hot coolant leaving the reactor core was designed to be 648°C. In order to obtain these high temperatures, yet ensure useful graphite core life (graphite oxidises readily in

CO_2 at high temperature) a re-entrant flow of coolant at the lower boiler outlet temperature of 278°C is utilised to cool the graphite, ensuring that the graphite temperatures do not vary too much from those seen in a Magnox station. The superheater outlet temperature and pressure were designed to be 170bar and 543°C.

The fuel is uranium oxide pellets, enriched to 2.5-3.5%, in stainless steel tubes. The original design concept of the AGR was to use a beryllium based cladding. When this proved unsuitable, the enrichment level of the fuel was raised to allow for the higher neutron capture losses of stainless steel cladding. This significantly increased the cost of the power produced by an AGR. The carbon dioxide coolant circulates through the core, reaching 640°C and a pressure of around 40 bar, and then passes through boiler (steam generator) assemblies outside the core but still within the steel lined, reinforced concrete pressure vessel. Control rods penetrate the graphite moderator and a secondary system involves injecting nitrogen into the coolant to hold the reactor down. A tertiary shutdown system which operates by injecting boron balls into the reactor has been proposed 'as retrofit to satisfy the Nuclear Installations Inspectorate's concerns about core integrity and core restraint integrity'.

The AGR was designed to have a high thermal efficiency (electricity generated/heat generated ratio) of about 41%, which is better than modern PWRs which have a typical thermal efficiency of 34%. This is due to the higher coolant outlet temperature of about 640 °C practical with gas cooling, compared to about 325 °C for PWRs.

AGRs are designed to be refueled without being shut down first. This on-load refuelling was an important part of the economic case for choosing the AGR over other reactor types, and in 1965 allowed the Central Electricity generating Board (CEGB) and the UK government to claim that the AGR would produce electricity cheaper than the best coal-fired power stations. However fuel assembly vibration problems arose during on-load refuelling at full power, so in 1988 full power refuelling was suspended until the mid-1990s, when further trials led to a fuel rod becoming stuck in a reactor core. Only refuelling at part load or when shut down is now undertaken at AGRs.

The AGR was intended to be a superior British alternative to American light water reactor designs. It was promoted as a development of the operationally (if not economically) successful Magnox design, and was chosen from a plethora of competing British alternatives There were great hopes for the AGR design. An ambitious construction programme of five twin reactor stations was quickly rolled out. The lead station, Dungeness B was ordered in 1965 with a target completion date of 1970. After problems with nearly every aspect of the reactor design it finally began generating electricity in 1983, 13 years late. The follow on stations all experienced similar problems and delays. The financing cost of the capital expended, and the cost of providing replacement electricity during the delays, were enormous, totally invalidating the pre-construction economic case. Currently there are seven nuclear generating stations each with two operating AGRs in the UK.

SUMMARY

The gas cooled reactors were one of the first entrants in the nuclear reactor arena and were developed in UK and France. However, problems with cladding corrosion and increased costs consequent to design improvements, made them less competitive to PWR and BWR. However, there is an interest in Gas Cooled Fast reactors and Very High Temperature gas cooled thermal reactors as a part of the Generation IV reactor concepts and a gas cooled fast reactor is planned to be built in France.

BIBLIOGRAPHY

1. Advanced Gas-cooled Reactors,Part 1 Volume 24, Issue 5, Nuclear Energy, BNES, 1985.
2. MAGNOX reactors, Volume 20, Issue 2, Nuclear Energy, BNES, 1981.
3. D N Simister, M Williams, Extended generation at the WYLFA magnox nuclear power station: An operational and regulatory perspective - HM Nuclear Installations Inspectorate, (2010).
4. J.Wood, Nuclear Power, Vol 52, IET Power and Energy Series, Institution of Electrical Engineers, (2006).
5. Health and Safety Executive(HSE), Magnox Nuclear Power Reactor Programme, (1994).
6. Gareth B Neighbour, Management of Ageing in Graphite Reactor Cores, Vol 309,Special Publication, Royal Society of Chemistry, Great Britain, (2007).
7. Gareth B Neighbour, Securing the safe performance of graphite Reactor Cores, Royal Society of Chemistry, Great Britain, (2010).
8. S. Glasstone and A. Sesonske, Nuclear Reactor Engineering, 3rd Edition, Von Nostrand (1981).
9. Kenneth D. Kok (ed.),Handbook of Nuclear Engineering, CRC Press, (2009).
10. Dan Gabriel Cacuci (ed.), Handbook of Nuclear Engineering,Springer (2010).
11. Steven B.,Krivit,Jay H.Lehr, Thomas B. Kingery , Nuclear Energy Encyclopedia: Science, Technology, and Applications, John Wiley (2011).

ASSIGNMENTS

1. What are the important characteristics of MAGNOX type reactors.
2. What are important problems faced in the operation of MAGNOX reactors.
3. What is the difference between MAGNOX and AGR.
4. Explain the operation of a MAGNOX reactor with a schematic diagram.
5. What led to the poor market for AGRs.

11

Liquid Metal Fast Reactors

11.0 INTRODUCTION

Natural uranium contains small quantity of U-235(0.7%), but large quantities of U-238 (99.3%). In slow (neutron) reactors like BWR, PWR and AGR, the scarce U-235 is used as fuel for power generation while the abundant U-238 is largely unused. In a fast (neutron) reactor, U-238 can be converted to a nuclear fuel (Pu-239). Thus a fast reactor helps extract large amount of power from the abundant U-238 (through conversion). The conversion is known as fuel breeding. Fig 11.1 gives the process of this conversion.

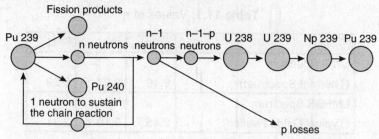

For breeding to be possible, the number of Pu_{239} nuclei produced must exceed the number of Pu_{239} nuclei consumed.
The diagram shows that for one plutonium nucleus consumed, n–p–1 are produced.
It is consequently necessary that n–1–p>1 and n>2+p>2.

Fig. 11.1: Condition for breeding

By proper choice of fuel materials and design of reactor, it is possible to produce more fissile material than consumed. For every neutron absorbed, 1 neutron is needed for sustaining chain reaction, 1 neutron to be captured by fertile U 238 nuclei to get converted to Pu 239. Allowing for other loss and parasitic absorption in structures, coolant etc. more than 2 neutrons are needed ($\eta>2$) to continue with the chain reaction and breed fuel.

Figure 11.2 shows that for U–Pu fuel, a sufficient value of η can be obtained only with a fast neutron spectrum. One notices that the condition for breeding is well fulfilled in fast spectrum reactor in particular for 239Pu for which the value of η averaged is about 2.45. The Table 11.1 provides values of η for U235 and Pu239 compared parameters in thermal and fast spectrum. In a fast neutron reactor, the neutrons available for conversion are more than in a thermal neutron reactor.. In such a reactor, there is net production of fissile material that can be used to fuel another reactor.

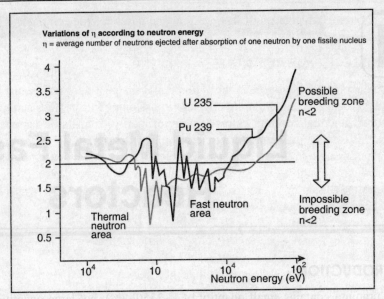

Fig. 11.2: Variation of η of Different Fuels

In a thermal reactor spectrum, for a U-233/Th-232 core η = 2.29 is sufficiently above 2.0, so that would produce more U-233 then it consumes. The values of η for both U-235 and Pu-239 are also above 2.0, but not so much above it that breeding is practical. So, at thermal energies, only the U-233/Th-232 cycle is possible for breeding.

Table 11.1: Values of η

	^{239}Pu	^{235}U	^{233}U
LWR Spectrum (Thermal Spectrum)	2.10	2.07	2.29
LMFBR Spectrum (Typical Oxide Fuelled)	2.45	2.10	2.31

There is no fertile nuclide that transmutes to U-235 upon neutron absorption. So, once supplies of U-235 have been consumed, it is gone. We can have a U-233/Th-232 cycle in thermal reactors or a Pu-239/U-238 cycle in a fast reactor.

The pressure resulting from a growing demand and linked to strong constraints on fossil fuels (USA and China's heavy dependence on coal; USA and Europe's dependence on oil and gas) should lead to the relaunching of nuclear plant construction in countries possessing nuclear technology and to the development of this same technology in others. The unavoidable steady reduction of economically exploitable resources of natural uranium will contribute to reliance on nuclear energy for sustainable development and cause nuclear energy to play a heightened role in the future. Only through the exploitation of fertile material and, therefore, through the use of fast neutron reactors can this be implemented. With fast neutron reactors, the reserves 238U, currently stored as tails (tails represent the depleted 238U, which remains after completion of the enrichment process) from enrichment plants alone, allows for increased energy reserves up to a factor of about 50 in comparison to the current light water reactor technology. Therefore, fast reactors are the key to an efficient use of uranium resources. This

Liquid Metal Fast Reactors

chapter brings out the advantages of fast reactors and describes the components of a Sodium cooled Fast Reactor (SFR).

11.1 FLEXIBLE USE OF ACTINIDES

Fast reactors play a unique role in the actinide management mission because they operate with high energy neutrons that are more effective in fissioning transuranics actinides (Pu, Am, Cm). In contrast, thermal reactors extract energy primarily from fissile isotopes; a thermal spectrum also leads to the generation of higher actinides that complicate subsequent recycling. Fast reactors can operate in three distinct fuel cycle roles. A conversion ratio (the conversion ratio is defined as the ratio of the actinide production rate to the destruction rate, whereas the breeding ratio is a similar ratio for the fissile material) less than 1 ("transmuter" mode) means that there is a net consumption of transuranics. Here, "transmute" means to convert transuranics into shorter-lived isotopes in order to reduce long-term waste management burdens. A conversion ratio close to 1 ("converter mode") provides a balance in transuranic production and consumption. This mode results in low reactivity loss rates with associated control benefits. A conversion ratio greater than 1 ("breeder mode") means that there is a net creation of transuranics. This approach allows the creation of additional fissile materials, but requires the inclusion of extra uranium in the fuel cycle. An appropriately designed fast reactor has the flexibility to shift between these operating modes and the desired actinide management strategy will depend on a balance of waste management and resource extension considerations.

Nuclides with even number of neutrons have very small cross section for neutron energies below 1 MeV. Hence in the thermal spectrum of light or heavy water reactors, these nuclides tend to accumulate. Consequently a continued burnup of LWR/PHWR demands more than 2 neutrons to transmute Np 237, Pu 240, Pu 242 and Am 241 into fission products Especially troublesome are the two latter isotopes, which need close to four and three neutrons respectively, in order to go through transmutation chains.

$$242Pu + n \rightarrow 243Pu \rightarrow 243Am + e^-$$
$$243Am + n \rightarrow 244Am \rightarrow 244Cm + e^-$$
$$244Cm + n \rightarrow 245Cm$$
$$245Cm + n \rightarrow 3.5n + 2FP + 200MeV$$

And

$$241Am + n \rightarrow 242Am + e^- \rightarrow 238Pu$$
$$238Pu + n \rightarrow 239Pu$$
$$239Pu + n \rightarrow 3.2n + 2FP + 200MeV$$

Fig 11.3 shows the ratio of fission to absorption (fission+ capture) cross sections in oxide fuelled LWR and FBR cores for transuranics nuclides of concern.

11.2 WASTE MINIMIZATION

Owing to their high level of neutron flux (10 times higher than that of a PWR) and to the ratios of capture cross sections on fission cross sections of major actinides more favorable in the fast spectrum (by a factor of 10), FBR's are the best tool for the transmutation of actinides in reactor. Two modes of waste incineration have been considered:
- The homogeneous mode wherein actinides to be transmuted (Np, Am, Cm) are mixed with standard fuel in limited quantities (a few percent), thereby "diluting" the impact on the reactor and the fuel cycle facilities.

- The mixed mode in which the elements are separated from the fuel and placed in a limited number of devices is known as "targets." The impact on all posts in the fuel cycle is then increased substantially but for a reduced flow of materials.

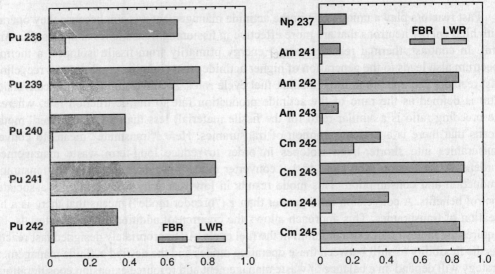

Fig. 11.3: Fission to absorption cross section ratio for Pu isotopes and minor actinides

In order to establish the scientific feasibility of these different modes, radiation experiments have been carried out in France in the PHENIX reactor. In conjunction with the actinide management, the FR technology offers the means to reduce waste generation by features such as improved thermal efficiency, greater use of fuel resources, and the development of superior waste forms for the FR closed fuel cycle. It is also obvious that FR nuclear power contributes to the reduction of the greenhouse effect (CO_2 emissions) compared to electricity generation using fossil fuels. Efforts will also be made to achieve reductions in the amount of waste generated from the operations and maintenance and the decommissioning of system facilities and the amount of waste migrating into the environment.

There are potential benefits of a closed fuel cycle based on fast reactors for waste management. It is easier to transmute Transuranics (TRU) or Minor Actinides (MA) in a fast reactor core, and there is less impact on the fuel cycle (e.g., at fuel fabrication). It is then possible to have a sustainable close cycle, with reduced burden on a deep geological storage. Certain elements (plutonium, americium, cesium, strontium, and curium) are primarily responsible for the decay heat that can cause repository temperature limits to be reached. Large gains in repository space are possible by processing spent nuclear fuel to remove those elements. Related to treatment of recovered elements, cesium and strontium can be stored separately for 200–300 years and plutonium, americium, and curium can be recycled for transmutation and/or fission by irradiation in fast reactors.

11.3 OVERVIEW OF SODIUM COOLED FAST REACTOR (SFR)

Tables 11.2 lists the existing experimental FBRs in the world. These reactors had essentially twofold aims: On the one hand, their purpose was to provide experience in the operation of sodium-cooled reactors on a sufficient scale, and on the other, they were designed to allow the development of a fuel element capable of withstanding high burnup. The next phase is that of demonstration reactors. These reactors are electricity generators with a power ranging between

100 and 600MWe, and their purpose is to prepare for the introduction of high-power reactors by validating their concepts. Table 11.3 lists these types of reactors in the world.

Table 11.2: EXPERIMENTAL REACTORS

	NAME	THERMAL POWER MW	CRITICALITY	COUNTRY	STATUS
PRELIMINARY	Clementine	0.02	1946	USA	Stopped 1952
	EBR-I	1.4	1941	USA	Stopped 1963
	BR-1		1955	RUSSIA	Stopped 1957
	BR-2	0.2	1956	RUSSIA	Stopped 2002
	BR-5/BR-10	5/10	1958/1973	RUSSIA	Stopped 1965
	LAMPRE	1	1961	USA	
EXPERIMENTAL	DFR	75	1959	UK	Stopped 1977
	EBR II	60	1963	USA	Stopped 1993
	EFFBR	200	1963	USA	Stopped 1972
	RAPSODIE	24/40	1967/1970	FRANCE	Stopped 1983
	BOR 60	60	1968	RUSSIA	Running
	SEFOR	20	1969	USA	Stopped 1972
	KNK I-KNK 2	60	1972/1977	GERMANY	Stopped 1991
	JOYO	50	1977	JAPAN	RUNNING
	FFTF	400	1980	USA	Stopped 1992
	FBTR	40	1985	INDIA	RUNNING
	CEFR	60	2010	CHINA	RUNNING

Table 11.3: DEMONSTRATION REACTORS

REACTOR	POWER MWe	CRITICALITY	COUNTRY	STATUS
EFFBR	100	1963	USA	Stopped 1972
BN-350	150	1972	KAZHAKASTAN	Stopped 1993
PHENIX	250	1973	FRANCE	Stopped 2009
PFR	250	1973	UK	Stopped 1994
BN 600	600	1980	RUSSIA	RUNNING
SNR 300	300		GERMANY	GIVEN UP
MONJU	280	1992	JAPAN	To Restart
CRBRP	350		USA	GIVEN UP
PFBR	500		INDIA	UNDER CONSTRUCTION

The last phase involves the construction of high-power prototype reactors, between 750 and 1000 MWe, in order to achieve economic competitiveness. The first such project was the Super-Phenix reactor designed and built by a European consortium composed of France, Germany, and Italy. Other projects undertaken include the DFBR in Japan, the ALMR in USA, the SNR2 in Germany, the CDFR in UK, and then the EFR again within the framework of a European consortium, but these projects have all been abandoned. The reasons for that are not technical, but they are mainly due to a rise of hostility to nuclear energy rise in western nations and a lack of economic competitiveness demanding the development of fast reactor technology. Today only the BN600, in Russia is in operation. However, in recent years we have witnessed a renewed interest in nuclear reactors for the sake of saving natural resources. This corresponds to the launching of the Generation-IV International Forum. Such an event has led several countries to boost research on fast reactors, not only in USA, Japan, and France but also in Europe under the aegis of the European Commission, in Korea, and in China. It should be noted that in India, this technology has never been abandoned.

11.4 LAYOUT OF SFR COMPONENTS

The SFR heat transport system (Fig 11.4) consists of primary, secondary, and steam–water circuits. The purpose of primary circuit is to transport the heat energy, generated in the core to the intermediate heat exchangers (IHX), from which the heat is transported to the secondary circuit. The secondary circuit, in turn transfers the heat to the water circulating in steam generators, to generate steam. The steam–water circuit is the conventional electricity generating system, called as balance of plant (BOP). There are two types of layout in SFR the Pool type and the loop type. In the loop type (Fig 11.4 A) the components of the primary sodium circuit namely the reactor vessel (RV), the Intermediate Heat Exchanger (IHX) and primary sodium pump (PSP) are arranged as separate units and linked through double walled pipes. In the pool type (Fig 11.4B), the main components in the primary circuit are main vessel (MV), also called reactor vessel (RV), which houses the core and contains the coolant, grid plate, primary sodium pumps (PSP), Intermediate Heat Exchanger (IHX) and top shield. The inner vessel separates the hot and cold sodium in the main vessel. The grid plate supports the core and facilitates circulating the sodium to the core subassemblies to remove the nuclear heat. Primary pumps provide required pressure head to the sodium to flow through the core subassemblies. The inner vessel helps in routing the hot sodium to the IHX without any pipes. The MV is surrounded by a Safety vessel (SV) which ensures that the chances of loss of coolant in a pool type is very remote. The IHX, steam generators (SG), and secondary sodium pumps (SSP) are the main components in the secondary circuit, in both layouts.

Apart from the above mentioned circuits, decay heat circuit consisting of dedicated heat exchangers immersed in hot pool (in the case of pool concept) and air heat exchangers to transport the heat to the atmosphere (ultimate heat sink) during decay heat removal operations are important systems from point of view of safety. Depending upon the system, this circuit consists of decay heat exchangers (DHR) and pumps.

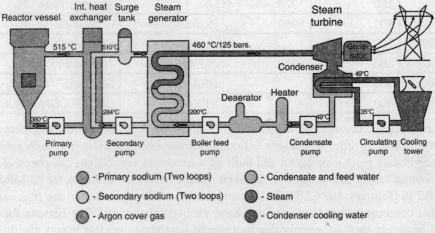

Fig. 11.4A: Loop Type SFR

Liquid Metal Fast Reactors

Compared to thermal reactors, SFR system operates with high operating temperature. Table 11.4 provides the core outlet and steam temperatures in a few SFR power plants. In view of higher steam temperatures, thermodynamic efficiency is around 40%, which implies less thermal pollution and less loss of nuclear heat. This is one of the main advantages of SFR plants compared to water reactors.

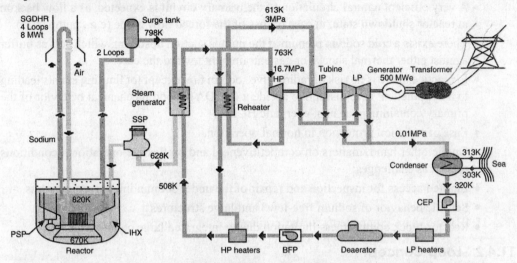

Fig. 11.4B: Pool Type SFR steam generators (SG), and secondary sodium pumps (SSP) are the main components in the secondary circuit.

Table 11.4: TEMPERATURES IN SFR

REACTOR	THERMAL POWER(MW)	ELECTRIC POWER(MW)	CORE OUTLET TEMPERATURE -K(°C)	STEAM TEMPERATURE -K(°C)
BOR 60	55	12	803 (530)	703 (430)
FBTR	40	13	788 (515)	753 (480)
EBR II	62.5	20	646 (473)	706 (433)
KNK II	58	21	798 (525)	758 (485)
PHENIX	563	250	833 (560)	785 (512)
PFR	650	270	833 (560)	788 (515)
MONJU	714	280	802 (529)	756 (483)
BN 350	750	350	710 (437)	678 (405)
BN 600	1470	600	823 (550)	778 (505)
SUPER-PHENIX	3000	1240	818 (545)	763 (490)
PFBR	1250	500	818 (545)	763 (490)

11.4.1 Pool Concept

The points of motivation for a pool-type primary circuit can be perceived as opposed to the loop concept:

- There is no relevant accident scenario of loss of primary coolant. The primary sodium inventory is managed by safety provisions (e.g., safety vessel)

- The large thermal inertia of the reactor block contributes to slow down any transient loss of heat sink.
- There is no risk to break the hydraulic loop from the core outlet toward the core inlet as all pipe entries are from the top.
- A very efficient natural circulation of the primary circuit is expected, as a flow backup at reactor shutdown state, in case of loss of the forced flow mode (e.g., pumps trip).
- There exists a cold sodium plenum at the pumps suction upstream, which acts as buffer against either thermal shocks or gas entrainment toward the core.
- In practice, there is no risk of radioactive sodium fire, except for limiting events leading to a hypothetical core disruptive accident (HCDA). Good mechanical behavior of the primary containment against energetic HCDA.
- Ease of radiation protection in normal operation.

On the other hand, matters on competitiveness and on flexible operational conditions remain as challenges:

- Limited access for inspection and repair of the under-sodium internal equipments
- Seismic behavior of sodium free-level and large structures
- Reactor-block compactness limitation due to integrated large components

11.4.2 Loop Concept

In the same way, motivation for loop concept and challenge as opposed to the pool concept can be perceived:

- Ease of access for maintenance and repair of those of the primary components located outside the reactor-block (e.g., IHX)
- Compactness of the reactor-block (e.g., vessel diameter) and reduction of the primary loops number
- Potential for further construction cost reduction and for innovative change of primary and inter-mediate equipment (e.g., pump design, integrated components, intermediate circuit change ...)
- Any rotating pump-shaft is away from the core vicinity

In return, the designer has to solve some issues about:

- Prevention of loss of the primary coolant (e.g., pipe integrity) and provisions to keep a hydraulic loop through the core whatever the abnormal operating conditions are
- Potential consequences of a LOCA, inside (e.g., gas entrainment) and outside the primary circuit (e.g., active sodium fire)
- Provision against asymmetric operating faults (e.g., trip of one of the pumps leading to reverse flow)
- Suitable implementation of several decay heat removal (DHR) systems in order to cope with different accidental configurations of the primary circuit
- Additional provisions regarding the radiological protection

11.5 FUEL DESIGN

11.5.1 Fuel Element

In most cases, the fuel element is a cylindrical solid of revolution. A circular section is the most adapted to allow the cladding to withstand the primary pressure stresses due to the

Liquid Metal Fast Reactors

fission gas release. Typical fuel element is shown in fig 11.5. The fuel pin is mainly made up of the cladding containing the fuel pellets or rods and is closed by two welded end plugs, which results in a gastight assembly. The clad material is austenitic stainless steel unlike zirconium in thermal reactors. This is because in the fast neutron spectrum, absorption cross sections are less. In the top part of the pin, there is generally a spring, which maintains the fuel column in place and creates a gap (plenum) for fission gas expansion and fuel elongation. Disks made of refractory materials (natural or depleted uranium oxide) are placed at the top and bottom parts of the fuel to insulate it from the metallic parts. The fuel pin can include columns made of fertile material making up the axial blankets.

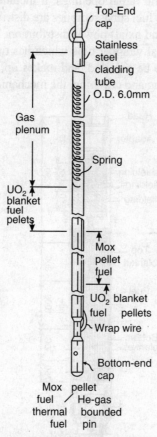

Fig. 11.5: Fuel Element of SFR

To improve heat transfer between fuel and cladding, the fuel element (Oxide) is filled with helium at atmospheric pressure. In the case of a fuel material that swells (metal alloy, carbide, and nitride), a large gap is needed, and the thermal gradient increases in the gap. To reduce it, one can fill the pin with sodium (possible because of the good compatibility between these materials), in order to obtain very good heat transfer between fuel and cladding and thus take full advantage of the good thermal conductivity of the fuel. In this case, an alternative solution would be to have a fuel element with a vent, an opening equipped with a filter allowing the fission gases to be continuously released from the pin.

A spacer wire, made of steel of the same nature as that of the cladding, is helically wrapped around each pin and attached to the end plugs. This solution
- Ensures regular spacing of the pins over the whole length

- Minimizes vibrations
- Ensures sodium mixing

11.5.2 Fuel Subassembly

For the purpose of illustration, a typical fuel subassembly is shown in Fig 11.6. Fuel subassembly design, as far as the number, diameter, and arrangement of fuel elements is concerned, is mainly guided by thermo mechanical, hydraulic, and neutronics considerations. The bottom nozzle of the subassembly allows it to be positioned in a diagrid which is a cold sodium collector and supplies it with sodium several radial openings. It includes a pressure drop device used to adjust the sodium flow rate, as the fuel subassemblies are distributed over several flow rate areas, varying according to the radial (and axial) power distribution. The median part includes the fuel pins contained in a steel hexagonal wrapper tube which has two functions: it forms the cooling channel, allowing sodium flow to be adjusted, and makes up, with the top and bottom nozzles to which it is connected by welding and crimping, the mechanical structure of the subassembly.

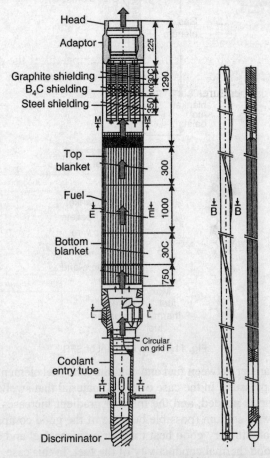

Fig. 11.6: Fuel Subassembly

At about the same level as the top of the fuel, the wrapper tube includes spacer pads, which are dimples impressed on each face of the tube, with a small gap under cold conditions between two neighboring subassemblies, in order to obtain a comp pact core in normal operation under effect of thermal expansion. The top nozzle, consisting of a steel hexagonal block with a circular central hole allowing circulation of sodium leaving the subassembly, is used both as neutron protection of the top components of the reactor and as the gripper head of the assembly for handling.

The FR fuel subassembly is characterized by:
- A high plutonium content (from 15 to 30%)
- A fuel in the shape of solid or annular pellets oof (U,Pu)O2, (U,Pu)N, (U,Pu)C or in the shape of rod of U–Pu–Zr metallic alloy
- A spacing made by a wire helically wound around each of pins (even though different solutions, such as grid spacing have been tested)
- A hexagonal wrapper tube
- The use of austenitic, ferritic, martensitic steels or nickel alloys as materials for structural components.

The reactor core of a SFR is depicted in Fig 11.7.

11.5.2.1 Fuel Handling

Fuel handling is done when the reactor is shutdown. There is sufficient reactivity built in the core so that frequent fuel loading is not required, unlike a PHWR. The fuel handling is done at a sodium temperature of 180 to 250°C. At lower temperatures the sodium aerosols in the argon cover gas deposit on the top shields and can prevent rotation of the top plugs. Also the top shields are kept at temperatures above 120°C, so that vapour condensate will not solidify on the upper parts.

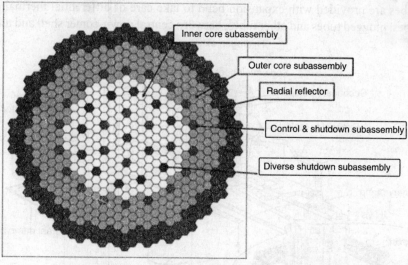

Fig. 11.7: SFR Core

11.6 INTERMEDIATE CIRCUITS

The intermediate/secondary sodium circuit is interposed between primary sodium circuit and tertiary steam–water system from considerations of reactor safety, in the sense that the primary sodium systems would be free from mechanical and chemical effects of sodium–water reaction, if it occurs in the Steam Generator(SG). The overall design incorporates provisions, for example, in vessel shielding around core in case of pool type layout, to render the secondary sodium circuit to be of nonradioactive, thus making its components generally amenable for maintenance with adequate care. Provisions such as surge tanks/cover gas in SG are incorporated in the circuit to absorb the pressure surges during inadvertent sodium–water reaction. Apart from piping, the important components of the circuit are IHX, secondary sodium pump (SSP), and SG (Fig 11.8). From an economic consideration, there is incentive to minimize the overall length of piping and optimize the location of various equipments, thereby leading to compact circuit layout. The operating experience of SFR worldwide has demonstrated the reliable operation of sodium pumps and IHXs as compared to steam generators, wherein maintenance operations on account of sodium–water reactions have led to prolonged shutdown of the respective plants. Therefore, plant availability considerations would desire incorporation of many SG modules.

11.6.1 Intermediate Heat Exchanger (IHX) (Fig 11.9)

IHX transfers thermal energy from radioactive sodium on primary side to the nonradioactive sodium on the secondary side. IHX along with the steam–water circuit will be the normal decay heat removal path. In a loop-type fast reactor, the primary coolant is circulated through IHX external to the reactor tank (but within the biological shield owing to the presence of radioactive sodium-24 in the primary coolant). In a pool type fast reactor, IHX is immersed in the reactor tank.

In IHX, tube bundle tubes can be either straight or with an expansion bend. Tube bundle with straight tubes has the advantages of lesser tube sheet thickness, ease of manufacture, less pressure drop, and low FIV due to the absence of bend. In some reactors (FBTR, PFR, BN-600, FFTF), tubes are provided with expansion bend to take care of differential thermal expansion among tubes, plugged tubes and other tubes, between central down-comer shell and tube bundle.

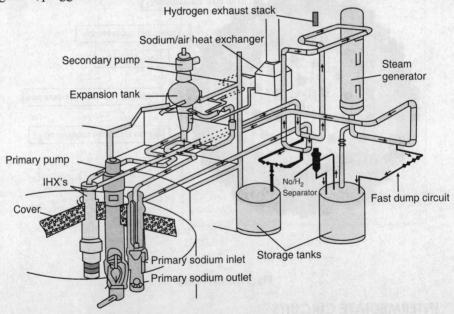

Fig. 11.8: Schematic of Secondary sodium circuit

All intermediate heat exchangers in pool-type reactors admit fluid on the shell side through circumferential openings in the shell, called windows. The sodium flow is governed by the difference in hot and cold pool levels, which should be equal to the pressure drop of primary sodium in IHX. Larger the pressure drop, the height of sodium in inner vessel would need to be higher. This would in turn raise the heights of main vessel, safety vessel, control plug etc. which increases the cost. Hence, it is paramount to ensure minimal pressure drop though IHX. Normally the pressure drop in pootype IHX is between 1 to 1.5 m of sodium. The shell of the heat exchanger can be rolled to full length and windows of required dimensions at appropriate elevations can be cut off.

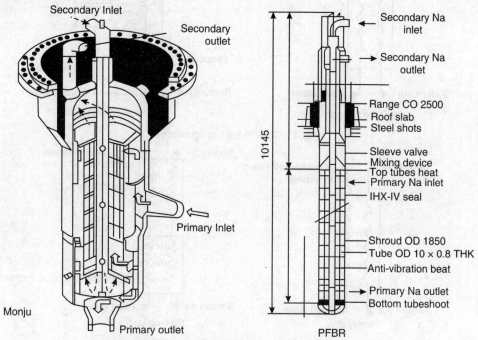

Fig. 11.9: IHX Configurations for LOOP and POOL type

Austenitic stainless steels are the principal materials preferred as structural material for IHX. Owing to its superior high temperature behavior, SS316 and SS304 are the most preferred stainless steels. Variants of these steels obtained through small modifications with their constitution such as L, LN, etc. are often used for specific requirements.

11.6.2 Steam Generators

Steam generator availability is recognized as the major factor for achieving good load factor, thereby commercial success of the SFR power plant. The reliability results not only from good design and fabrication practices but also from a proper selection of the structural material. A worldwide survey of fast reactor steam generators (SGs) in operation or under development shows considerable diversity in design and selection of materials. Many factors that are often difficult to quantify are involved, and considerable amount of judgment is required in assessing a steam generator design. The operating conditions of fast reactor SG are not severe compared to that of fossil fueled power plants; however, the tube leakage leading to severe sodium–water reaction affects the availability of the plant. No satisfactory design, which will ensure complete protection against the tube leakage, has been evolved. Due to this reason, no standard steam generator concept is available. Countries such as US, France, UK, Japan, and India have adopted diverse design for their fast reactor program.

The two different concepts generally followed for the steam generator design (Fig 11.10) are
- Recirculation type
- Once-through type

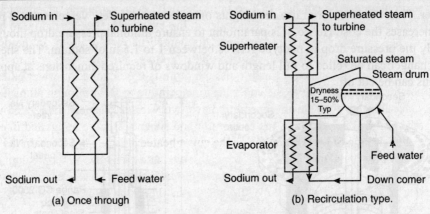

Fig. 11.10: SFR SG Configurations

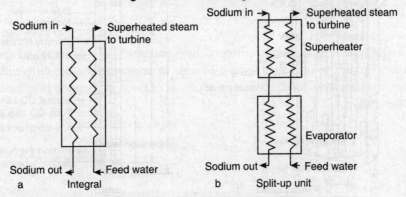

Fig. 11.11: Integral and Split up SG

Table 11.5: SG CONFIGURATIONS SFRs

REACTOR	SG TYPE	CONFIGURATION	MATERIAL
PHENIX	Once Through split up modular unit	Serpentine	2.25 Cr 1 Mo for evaporator SS 321 for superheater and reheater
FBTR	Once Through Integral modular unit	Serpentine	2.25 Cr 1 Mo stablished
PFR	Recirculation	U tube	2.25 Cr 1 Mo for evaporator SS 316 for superheater and reheater
BN 350	Recirculation	Bayonet tube	2.25 Cr 1 Mo
EBR II	Recirculation Duplex tube	Straight tube with cold springing shell	2.25 Cr 1 Mo
Monju	Once through split up unit	Helical	2.25 Cr 1 Mo for evaporator SS 321 for superheater
BN 600	Once Through Modular Units	Straight tube with bellows in shell	2.25 Cr 1 Mo for evaporator SS 304 for superheater
PFBR	Once Through Integral modular Units	Straight tube with expansion bend	Modified 9 Cr 1 Mo

Liquid Metal Fast Reactors

The recirculation type consists of two units, namely, evaporator and superheater, which are connected through a steam drum. The water in the evaporator gets heated up to a certain quality and then the water–steam mixture gets discharged into steam drum. In the drum, the steam separators separate steam from the two phase mixture, and the steam goes to the super heater for superheating. In once-through type, the drum is absent, and the water entering the steam generator gets heated up to the super heat temperature passing through all heat transfer regimes. Once-through type steam generators have two variants, namely, integrated unit and split up unit. In the integrated unit, the evaporator and super heater are integrated in one unit, and water enters to the SG and come out as the super heated steam (Fig 11.11). In the split up unit, the evaporator and super heater are separate units, and these two units are connected in series. Water enters to the evaporator and comes out as saturated steam, which enters to the super heater and come out as the super heated steam. The type of SG adopted by different countries is given in table. 11.5.

Fig 11.12 shows the SG configuration adopted in the Fast Breeder Test Reactor (FBTR) and Prototype Fast Breeder Reactors(PFBR) in India. All the new generation reactors have high power capacity and are now adopting once-through SGs. The SGs are shell and tube type, and different designs exist due to the different configurations of shell and tubes. Tube and shell configurations should be such that it must be able to accommodate the differential thermal expansion of the tube bundle and shell. The selection of the configuration is influenced by the following factors:

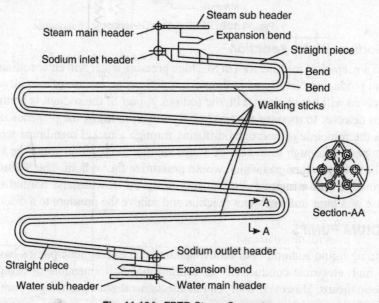

Fig. 11.12A: FBTR Steam Generator

- Good tube bundle accessibility for inspection, maintenance and repair
- Relative freedom from sodium stratification
- Performance sensitivity with plugged tubes
- Ability to limit damage in the event of a sodium–water reaction

Based on the above requirements, different designs are evolved in different countries in which tube configuration may be either straight tube, helical, serpentine, or U tube. Tube may be either single wall tube or duplex wall tube. Among various choices available, each country adopted different designs in the early stages of the LMFBR reactor program.

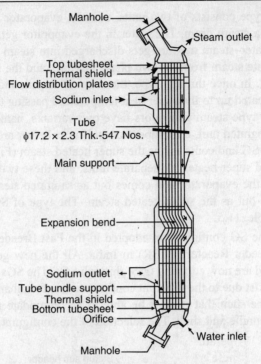

Fig. 11.12B: PFBR Steam Generator

11.6.2.1 Sodium Water Reaction

In case of any crack in the tube wall of SG, high pressure water will enter sodium and sodium water reaction producing NaOH and hydrogen gas will occur. In case of a small water leak the hydrogen evolved will travel along with the sodium. A part of the sodium is continuously sent to a hydrogen detector to monitor the level of Hydrogen to know the presence of a leak. The detector uses the principle of hydrogen diffusing through a nickel membrane when hydrogen laden sodium passes through it. In case of a tube rupture the reaction would be very fast with large generation of hydrogen gas which would pressurize the sodium. The Sodium headers in the SG is provided with a rupture disc suitably designed to withstand normal transients but rupture in case of a large sodium water reaction and relieve the pressure to a discharge circuit.

11.7 SODIUM PUMPS

For pumping liquid sodium, both centrifugal and electromagnetic pumps have been used. The relative high electrical conductivity of sodium makes it amenable to being pumped by electromagnetic means. However, with reliable mechanical seals centrifugal pumps have been used for main heat transport system.

11.7.1 Electromagnetic Pump

The operation of electromagnetic pumps is based on the principle that a conductor carrying current in a magnetic field experiences a force. The main advantage of these pumps are: No moving parts and absence of free liquid surface. The major disadvantage is the poor pump efficiency of 10-20% as compared to 85% in centrifugal pumps. The poor efficiency is due to IR losses in the sodium and duct (~50%). For larger sodium flows these losses come down. There have been conduction and induction pumps both operating in AC/DC modes. Figure 11.13 shows the schematic of a DC conduction pump.

Liquid Metal Fast Reactors

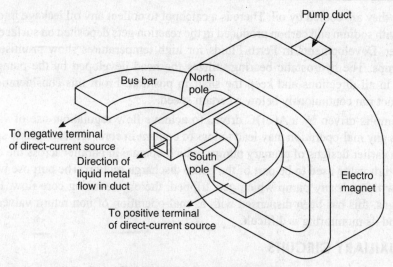

Fig. 11.13: DC Conduction Pump

11.7.2 Centrifugal Pumps (Fig 11.14)

Sodium pumps are generally free surface, single stage, centrifugal units. Since these pumps would operate at temperatures or 400 °C, ring wear clearences are increased compared to low temperature pumps. These are vertical pumps with a mechanical the top between shaft and

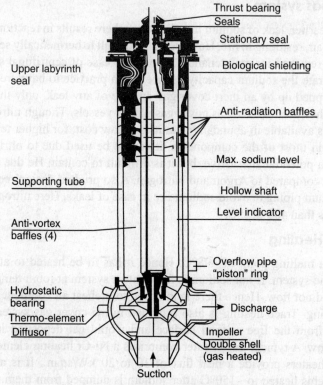

Fig. 11.14: Centrifugal sodium Pump

casing and a hydrostatic bearing to guide the shaft at the bottom. Hydraulic design of sodium pumps is akin to conventional water pumps. Since the mechanical seals at the top are in contact with argon, they are cooled by oil. There is a catchpot to collect any oil leakage into the pump. Oil reacts with sodium and carbon produced in the reaction gets deposited on surfaces impeding heat transfer. Developments in Ferro Fluids for high temperatures show promise for use in sodium pumps. The hydrostatic bearing utilizes the head developed by the pump to ensure equal flow in all directions and keep the shaft in position. From this consideration, sodium pumps are not run continuously below a certain speed.

The pump is driven by a AC/DC drive. To achieve flow regulation, use of valves is not preferred as any mal-operation may lead to loss of coolant in core. Hence variable speed drives are used. In earlier designs of primary pumps, which operate in parallel across the core, a float type non return valve used to be part of the pump discharge section. The purpose was to avoid reverse flow through any pump which gets tripped, thereby reducing core flow. However in future designs, this has been dispensed with as mal-operation of non return valves cannot be ruled out and its monitoring is difficult.

11.8 AUXILIARY CIRCUITS

Many auxiliary systems are needed for successful operation of the sodium cooled fast reactors plants. Many of these are large systems like heating and ventilation, compressed air etc., but these are not unique to Fast Reactors. In the following, the auxiliary systems unique to sodium cooled reactors would be described briefly. These are:1) Inert gas, 2) Trace Heating and 3) Sodium Purification.

11.8.1 Inert gas system

As indicated earlier, leak of sodium into the atmosphere results in reaction between sodium and oxygen in the air, resulting in fire. However it is difficult to hermetically seal all components as intervention is required for maintenance. Also in the case of centrifugal sodium pump, the shaft has to penetrate the sodium capacity. Hence it is a practice to have free sodium level in such capacities topped up by an inert cover gas. In case of any leak, only the inert gas would move out. Argon has been selected as cover gas within vessels. Though nitrogen satisfies this requirement and is available in abundant quantities at low cost, for higher temperatures > 400 deg, C that exist in most of the components, it cannot be used due to nitriding problems in steels. Helium is a potential alternative, but it is difficult to contain He due to its low density and is also costly compared to Argon and Nitrogen, Also prudent design requires inert gas in cells housing sodium piping to avoid sodium fire in case of leaks. Here nitrogen is preferred as temperature is less than 400 deg. C.

11.8.2 Trace Heating

Sodium has a melting point of ~98°C. Hence it has to be heated to at least 120-150°C before filling in the system. If hot sodium is filled into a system at room temperature, it would solidify and would not flow. Hence there is the need to preheat all components using sodium, including the piping. Trace heating is also provided in the cover gas systems as these carry sodium aerosols from the free sodium surface and would get deposited in cooler portions obstructing gas flow. A typical trace heater comprises a Ni-Cr heating element insulated with Magnesia. Such heaters provide a heat flux of 10 to 20 kW/sq.m.. It is also imperative to keep sodium systems heated to ~150°C after sodium is dumped from them, during shutdown conditions for maintenance.

11.8.3 Sodium purification system

The use of sodium as a coolant in fast breeder reactors needs purification of commercial sodium to nuclear grade (Table 11.6). The most important impurities in commercial sodium contributing to corrosion are oxygen and hydrogen. In addition there are other impurities like calcium and carbon. It is the impurities such as O, H and C which are of greater concern in the purification and handling of alkali metals. Because of the reactive nature of these metals, these impurities will ingress into sodium from air, moisture, Carbon-di-oxide, oils, and greases. Oxygen would thus be present not only as dissolved oxide, but also as a surface coating (solid oxide). The case of hydrogen is similar. Even though carbon solubility is low, carbon can be present as dispersed particulates.

The main techniques used in alkali metal purification are filtration, and cold trapping. Both filtration and cold trapping take advantage of the low solubility of many elements, in alkali metals just above the melting point. The impurities are filtered off in the initial charging stage while cold trapping is used as online purification. The other method is oxide slagging technique where oxides of many metals, such as calcium, are much stable than the oxide of the alkali metal being purified. Reaction with controlled quantities of oxygen thus precipitates out such metal oxides which are then removed.

SUMMARY

This chapter has traced the advantages of a fast reactor from breeding and actinide transmutation considerations. The various components of a Sodium Cooled Fast Reactor have been briefly described alongwith their basic functions. Though there is a lull in the building of Fast Reactors in USA and Europe due to nontechnical reasons, the large demand for power has necessitated the continued development of such reactors in China and India.

BIBLIOGRAPHY

1. Y. S. Tang, R. D. Coffield, Jr., and R. A. Markley, Jr.Thermal Analysis of Liquid-Metal Fast Breeder Reactors, American Nuclear Society, (1978).
2. A.E.Waltar, A.Reynolds, Fast Breeder Reactors,Pergammon Press,(1981).
3. Alan E. Waltar, Donald R. Todd and Pavel V. Tsvetkov, Fast Spectrum Reactors, Springer; 2012 edition .
4. J.G.Yevick, Fast Reactor Technilogy-Plant Design,MIT Press(1966).
5. A.Judd, Fast Breeder Reactors, Pergammon Press, Oxford(1981).
6. Kenneth D. Kok (ed.),Handbook of Nuclear Engineering, CRC Press, (2009).
7. Dan Gabriel Cacuci (ed.), Handbook of Nuclear Engineering,Springer (2010).
8. Steven B.,Krivit,Jay H.Lehr, Thomas B. Kingery , Nuclear Energy Encyclopedia: Science, Technology, and Applications, John Wiley (2011).
9. S.C.Chetal, V. Balasubramanian, P. Chellapandi, P. Mohanakrishnan, P. Puthiyavinayagan, C.P. Pillai, S. Raghupathy, T,K, Shanmugam, C. Sivathanu Pillai, Nuclear Engineering and Design, 236(2006)852-860.
10. G.Srinivasan, K.V. Suresh Kumar,B. Rajendran,P.V. Ramalingam, The Fast Breeder Test Reactor- Design and operating experiences, Nuclear Engineering and Design, 236(2006)796-811..

ASSIGNMENTS

1. Describe the principle of operation of a Sodium cooled fast reactor.
2. What are advantages of Fast Reactors over thermal reactors.
3. Compare LOOP and POOL type fast reactors.
4. What is the need for a secondary sodium circuit in a sodium cooled fast reactor.
5. Why is the steam generator considered as a critical component of a fast reactor.
6. How is sodium purified.
7. How is sodium water reaction detected.
8. What is the principle of operation of electromagnetic pumps. What are their efficiencies compared to centrifugal sodium pumps.
9. What are the different tube configurations in the different steam generators of SFR. Why so many designs have evolved.
10. Collect information on FBTR and PFBR reactors from literature and present in a seminar.

12

Molten-Salt Reactors

12.0 INTRODUCTION

Investigation of Molten-Salt Reactors(MSR) started in the late 1940's as part of the United States' program to develop a nuclear powered airplane. A liquid fuel appeared to offer several advantages, so experiments to establish the feasibility of molten salt fuels were begun in 1947. Liquid molten fluoride salts were the main line of effort at the Oak Ridge National Laboratory. The fluorides appeared particularly appropriate because they have high solubility for uranium, are among the most stable of chemical compounds, have very low vapor pressure even at red heat, have reasonably good heat transfer properties, are not damaged by radiation, do not react violently with air or water, and are inert to some common structural metals. This chapter traces the history of development efforts on MSR.

12.1 ORNL DEVELOPMENTAL WORK

A small reactor, the Aircraft Reactor Experiment (ARE) was built at Oak Ridge to investigate the use of molten fluoride fuels for aircraft propulsion reactors and particularly to study the nuclear stability of the circulating fuel system. The ARE fuel salt was a mixture of NaF, ZrF_4, and UF_4, the moderator was BeO, and all the piping was Inconel. In 1954, the ARE was operated successfully for 9 days at steady-state outlet temperatures ranging up to 1580°F (1133 K) and at powers up to 2.5 MWt. No mechanical or chemical problems were encountered, and the reactor was found to be stable and self-regulating.

That molten-salt reactors might be attractive for civilian power applications was recognized from the beginning and in 1956 H.G. MacPherson formed a group to study the technical characteristics, nuclear performance, and economics of molten-salt converters and breeders. After considering a number of concepts over a period of several years, MacPherson and his associates concluded that graphite-moderated thermal reactors operating on a thorium fuel cycle would be the best molten-salt systems for producing economic power. The thorium fuel cycle with recycle of 233U was found to give better performance in a molten-salt thermal reactor than a uranium fuel cycle in which 238U is the fertile material and plutonium is produced and recycled.

Studies of fast spectrum molten-salt reactors indicated that good breeding ratios could be obtained, but very high power densities would be required to avoid excessive fissile inventories. Adequate power densities appeared difficult to achieve without going to novel and untested heat removal methods. Two types of graphite-moderated reactors were considered—single-fluid reactors in which thorium and uranium are contained in the same salt, and two-fluid reactors in which a fertile salt containing thorium is kept separate from the fissile salt which contains uranium. The two-fluid reactor had the advantage that it would operate as a breeder; however, the single-fluid reactor appeared simpler and seemed to offer low power costs, even though the breeding ratio would be below 1.0, using the technology of that time. The fluoride volatility process, which could remove uranium from fluoride salts, had already been demonstrated in the recovery of uranium from the ARE fuel and thus was available for partial processing of salts from either type of reactor. The results of the ORNL studies were considered by a US Atomic Energy Commission task force that made a comparative evaluation of fluid-fuel reactors early in 1959. One conclusion of the task force was that the molten-salt reactor, although limited in potential breeding gain, had "the highest probability of achieving technical feasibility."

12.2 MOLTEN SALT REACTOR EXPERIMENT (MSRE)

By 1960, more complete conceptual designs of molten-salt reactors had emerged. Although emphasis was placed on the two-fluid concept because of its better nuclear performance, the single-fluid reactor was also studied. ORNL concluded that either route would lead to low-power-cost reactors, and that proceeding to the breeder either directly or via the converter would achieve reactors with good fuel conservation characteristics. Since many of the features of civilian power reactors would differ from those of the ARE, and the ARE had been operated only a short period, another reactor experiment was needed to investigate some of the technology for power reactors. The design of the Molten-Salt Reactor Experiment (MSRE) was begun in 1960. A single-fluid reactor was selected that in its engineering features resembled a converter, but the fuel salt did not contain thorium and thus was similar to the fuel salt for a two-fluid breeder. The MSRE fuel salt is a mixture of uranium, lithium-7, beryllium, and zirconium fluorides. Unclad graphite served as the moderator (the salt does not wet graphite and will not penetrate into its pores if the pore sizes are small). All other parts of the system that contact salt were made from the nickel-base alloy, INOR-8 (also called Hastelloy-N), which was specially developed in the aircraft program for use with molten fluorides. The maximum power was ~8 MWt, and the heat was rejected to the atmosphere. Construction of the MSRE began in 1962, and the reactor was first critical in 1965. Sustained operation at full power began in December 1966.

Successful completion of a six-month run in March of 1968 brought to a close the first phase of operation during which the initial objectives were achieved. The molten fluoride fuel was used for many months at temperatures > 1200°F (920 K) without corrosive attack on the metal and graphite parts of the system. The reactor equipment operated reliably and the radioactive liquids and gases were contained safely. The fuel was completely stable. Xenon was removed rapidly from the salt. When necessary, radioactive equipment was repaired or replaced in reasonable time and without overexposing maintenance personnel.

The second phase of MSRE operation began in August 1968 when a small processing facility attached to the reactor was used to remove the original uranium by treating the fuel salt with fluorine gas. A charge of ^{233}U fuel was added to the same carrier salt, and on October 2 the MSRE was made critical on ^{233}U. Six days later the power was taken to 100 kW by Glenn T. Seaborg, Chairman of the US Atomic Energy Commission, bringing to power the first reactor to operate on ^{233}U.

During the years when the MSRE was being built and brought into operation, most of the development work on molten-salt reactors was in support of the MSRE. However, basic chemistry studies of molten fluoride salts continued throughout this period. One discovery during this time was that the lithium fluoride and beryllium fluoride in a fuel salt can be separated from rare earths by vacuum distillation at temperatures near 1000°C. This was a significant discovery, since it provided an inexpensive, on-site method for recovering these valuable materials. As a consequence, the study effort looking at future reactors focused on a two fluid breeder in which the fuel salt would be fluorinated to recover the uranium and distilled to separate the carrier salt from fission products. The blanket salt would be processed by fluorination alone, since few fission products would be generated in the blanket if the uranium concentration were kept low. Graphite tubes would be used in the core to keep the fuel and fertile streams from mixing. Analyses of these two-fluid systems showed that breeding ratios in the range of 1.07 to 1.08 could be obtained, which, along with low fuel inventories, would lead to good fuel utilization. In addition, the fuel cycle cost appeared to be quite low. Consequently, the development effort for future reactors was aimed mainly at the features of two fluid breeders. A review of the technology associated with such reactors was published in 1967. The major disadvantage of this two-fluid system was recognized as being that the graphite had to serve as a piping material in the core where it was exposed to very high neutron fluxes.

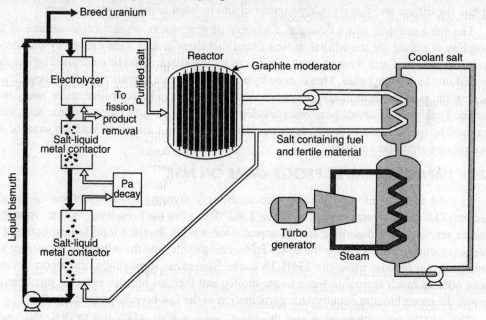

Fig. 12.1: Simplified flow diagram of a single-fluid MSBR

In late 1967, new experimental information and an advance in core design caused the molten-salt program at ORNL to change from the two-fluid breeder to a single-fluid breeder. Part of the information influencing this change concerned the behavior of graphite at higher radiation exposures than had been achieved previously, and the other part related too a development in chemical processing.

The irradiation data showed that the kind of graphite planned for use in an MSBR changes dimensions more rapidly than had been anticipated. This made it necessary to lower the core power density for the graphite to have an acceptable service life, and to plan on replacement of the core at fairly frequent intervals. Moreover, complexities in the assembly of the core seemed to require that the entire core and reactor vessel be replaced whenever a graphite element reached its radiation limit or developed a leak. Under such circumstances, many years of operation of a prototype reactor would be required to prove convincingly that the two-fluid core is practicable.

12.3 PROBLEMS OF MSR

Three problems requiring further development turned up during the construction and operation of the MSRE. The first was that the Hastelloy N used for the MSRE was subject to a kind of "radiation hardening," due to accumulation of helium at grain boundaries. Later, it was found that modified alloys that had fine carbide precipitates within the grains would hold the helium and restrain this migration to the grain boundaries. Nevertheless, it is still desirable to design well-blanketed reactors in which the exposure of the reactor vessel wall to fast neutron radiation is limited.

The second problem concerned the tritium produced by neutron reactions with lithium. At high temperatures the radioactive tritium, which is, of course, chemically like hydrogen, penetrates metals quite readily, and unless captured in some way, would appear in the steam generators and reach the atmosphere. After considerable development work, it was found that the intermediate salt coolant, a mixture of sodium fluoride and sodium fluoroborate, would capture the tritium and that it could be removed and isolated in the gas purge system.

The third problem came from the discovery of tiny cracks on the inside surface of the Hastelloy N piping for the MSRE. It was found that these cracks were caused by the fission product tellurium. Later work showed that this tellurium attack could be controlled by keeping the fuel on the reducing side. This is done by adjustment of the chemistry so that about 2% of the uranium is in the form of UF_3, as opposed to UF_4. This can be controlled rather easily now that good analytical methods have been developed. If the UF_3 to UF_4 ratio drops too low, it can be raised by the addition of some beryllium metal, which, as it dissolves, will rob some of the fluoride ions from the uranium.

12.4 IMPACT OF LMFBR PROGRAMME ON MSR

In 1962 the USAEC first took a position strongly favoring the development of breeder reactors. This position was spelled out in its 1962 "Report to the President." In that report two breeder reactors were described and discussed. One was the familiar liquid-metal-cooled fast breeder reactor (LMFBR) using the 238U-Pu breeding cycle, and the other was the molten salt fueled thermal breeder using the 233U-Th cycle. Somewhat surprisingly, the report devoted about 60% as much favorable space to the molten salt thermal breeder as to the fast breeder, despite the much broader countrywide participation in the fast breeder program.

The AEC's 1962 "Report to the President" was written while the MSRE was under construction, but apparently ORNL had proceeded far enough with the development program to impress the Subcommittee on Reactors of the General Advisory Committee with the value of the molten salt system. As soon as the "Report to the President" presented its justification for the development of breeder reactors, however, the existent Liquid Metal Cooled Reactors Department of the DRD proposed an elaborate and wide-ranging program for the development of the LMFBR concept, and that program began to gain momentum. On the other hand, those involved in the MSR program chose to wait until the MSRE had operated successfully before trying to expand the program. However by the time, the problems related to MSRE were found to be surmountable, the momentum and money needs of the LMFBR program were massive and there was no interest in funding a competitor.

SUMMARY

The MSRE was a very successful experiment, in that it answered many questions and posed but a few new ones. Perhaps the most important result was the conclusion that it was quite a practical reactor. It ran for long periods of time, and when maintenance was required, it was accomplished safely and without excessive delay.

Also, it demonstrated the expected flexibility and ease of handling the fuel. As mentioned above, it was the first reactor in the world to operate with 233U as the sole fuel, and the highly radioactive 233U used would have been extremely difficult to handle if it had had to be incorporated into solid fuel elements. In preparation for the run with 233U, the 235U was removed from the carrier salt in 4 days by the fluoride volatility process. It may be noted that MSR forms one of the concepts chosen by GEN IV for future development besides liquid metal fast breeder reactors.

BIBLIOGRAPHY

1. H.G.MacPherson, Molten Salt Reactor Status Report, University of Michigan Library (1958).
2. H.G. MacPherson, The Molten Salt Adventure", NUCLEAR SCIENCE AND ENGINEERING, Vol. 90, pgs 374-380 (1985).
3. The Development Status of MOLTEN-SALT BREEDER REACTORS", ORNL-4812, (1972).
4. U. Gat, J.R. Engel, & H.L. Dodds ,MOLTEN SALT REACTORS FOR BURNING DISMANTLED WEAPONS FUEL", Nuclear Technology (1992).
5. J.R. Engel, W.A. Rhoades, W.R. Grimes, and J.F. Dearing, Molten-Salt Reactors for Efficient Nuclear Fuel Utilization without Plutonium Separation", NUCLEAR TECHNOLOGY, Vol. 46, Nov (1979).
6. Kenneth D. Kok (ed.),Handbook of Nuclear Engineering, CRC Press, (2009).
7. Dan Gabriel Cacuci (ed.), Handbook of Nuclear Engineering,Springer (2010).
8. Steven B.,Krivit,Jay H.Lehr, Thomas B. Kingery , Nuclear Energy Encyclopedia: Science, Technology, and Applications, John Wiley (2011).

ASSIGNMENTS

1. What are the fuel and coolant in a MSR.
2. What are the advantages of MSR over other types of reactor.
3. Explain the work done at ORNL on MSR.

India's Nuclear Power Programme

13.0 INTRODUCTION

The beginning of the Indian nuclear program occurred prior to the independence of India from British rule. In 1943, Dr. Homi Jehangir Bhabha submitted a letter to the Sir Dorab Tata Trust to found a nuclear research institute. The approval of this proposal in April 1945 led to the creation of the Tata Institute of Fundamental Research (TIFR). TIFR began operations in Bangalore in June 1945 with Bhabha serving as the first director. In December 1945, Bhabha moved TIFR to Bombay and its official inauguration was on December 19, 1945. India became an independent state on August 15, 1947. A year after that, the government of India passed the Atomic Energy Act of 1948 leading to the establishment of the Indian Atomic Energy Commission (AEC), which would pursue in-depth studies on nuclear energy. At a press conference in Madras (presently called Chennai), Prime Minister Nehru spoke about the values of developing atomic energy, as follows "We are interested in atomic energy for social purposes. Atomic energy represents a tremendous power. If this power can be utilized as we use hydroelectric power, it will be a tremendous boon to mankind, because it is likely to be move the development from the social point of view." A four-year plan was unveiled to develop India's nuclear infrastructure for nuclear material exploration and the application of nuclear energy in medicine. During the period, Dr. Bhabha began seeking technical information on reactor theory, design, and technology from the U.S., Canada, and the U.K., while negotiating the sale or trade of raw materials such as monazite and beryllium-containing ore. In August 1950, Indian Rare Earths Limited was established for recovering minerals and the processing of rare-earths compounds and thorium-uranium concentrates. Later in April 1951, uranium deposits were discovered at Jaduguda, in the state of Bihar and drilling operations commenced in December 1951. The Jaduguda mine was the main source of uranium for the entire Indian nuclear program until the present day. This chapter is devoted to describe the setting up of the Atomic Energy Establishment at Trombay and further growth of the atomic energy in India in the power sector.

13.1 SETTING UP OF NUCLEAR ESTABLISHMENT

The Department of Atomic Energy (DAE) was established on August 3, 1954 with Dr. Bhabha as Secretary. Bhabha then conducted a series of meetings with British officials to request assistance in constructing a nuclear reactor and in converting uranium ores into metal for fabrication. He requested five tonnes of heavy water for use in a planned Indian research reactor. The British encouraged Bhabha to approach the Atomic Energy of Canada Limited (AECL) for the supply, as the U.K. had a deficit of heavy water available for domestic use.

On November 26, 1954, Bhabha outlined the three-stage nuclear energy plan for national development (Figure 12.1). Under this plan, India would start its nuclear energy generation in the first-stage with natural uranium-fueled, heavy-water moderated Pressurized Heavy Water Reactors (PHWRs) to produce power and plutonium. The first-stage reactors would be based on the CANDU technology and be built with Canadian assistance. These reactors were planned to generate 420 GWe-yrs of electricity in its lifetime.

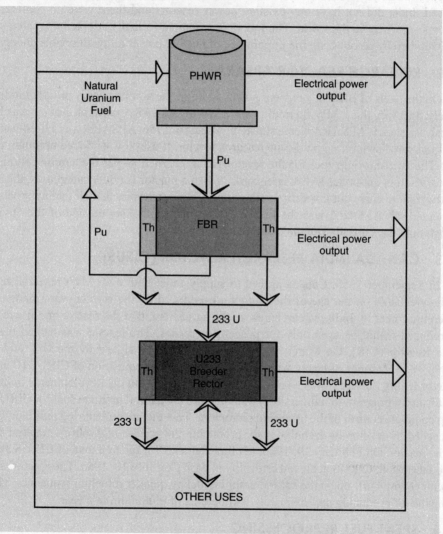

Fig. 13.1: Bhabhas 3 Stage Nuclear Power Plan

In the second-stage, plutonium separated from the spent fuel of the first-stage reactors would be used to power Fast Breeder Reactors (FBRs). The FBRs would then generate an additional 54,000 GWe-yrs of electricity. While generating power, the FBR's would also irradiate thorium in the blanket of the FBRs to breed U-233. In the third-stage, the U-233 bred from the second-stage would serve as fuel for the U-233 breeder reactors. These U-233 breeder reactors would further generate 358,000GWe-yrs of electricity and breed additional fissile materials. Effectively, a breeder reactor produces more fuel during operation than it consumes. The breeder reactor achieves this through a design that highly conserves neutrons in the system. These reactors would produce enough material to fuel themselves and produce additional fissile material (plutonium and/or U-233), which conceivably could be used either for weapons or to fuel power reactors.

The basis of three-stage-program was the indigenously available technology for production of natural uranium fuel assemblies, the vast reserves of thorium in India, and obtaining mastery of heavy water production and spent fuel reprocessing technology. When this program was devised, India did not have any existing power reactors and there were no commercial FBR systems anywhere in the world. The Indian government formally adopted this three-stage plan in 1958, thereby recognizing the importance of nuclear power as a sustainable energy source.

13.2 RESEARCH REACTOR APSARA

On the heels of Bhabha's nuclear power strategy, the atomic energy program in India grew rapidly. In 1955, the 1 MW thermal power (MWth) APSARA research reactor was built with British assistance. APSARA went critical on August 4, 1956. APSARA was a light-water cooled and moderated swimming pool-type research reactor. It uses low-enriched uranium, plate-type fuel. The British-origin fuel for the reactor is safeguarded as per the supply contract but the reactor itself is not under IAEA safeguards. When a reactor is under safeguards, the IAEA has got the right to carry out inspections from time to time to ensure that all fuel materials are fully accounted for. APSARA was the first operating reactor in Asia outside of the Soviet Union though only days ahead of Japan's first reactor.

13.3 CANADA INDIA RESEARCH REACTOR (CIRUS)

In September 1955, Canada agreed to supply India with a 40 MWt research reactor. No strict safeguards on the use of the plutonium produced by the reactor were made other than the commitment by India, via an annex to the agreement, that the reactor and fissile materials it produced would be used only for peaceful purposes. The reactor was named the Canada-India Reactor (CIR). On March 16, 1956, a contract was signed by the U.S. and India for supply of 18.9 tonnes of heavy water for the reactor. The acquisition of CIRUS (Canada India Reactor Utility Services) was a landmark event for atomic energy developments in India. India established a program to indigenously manufacture the natural uranium fuel for CIRUS so as to keep complete control of the produced plutonium. This fuel manufacturing program, eventually succeeded in developing techniques for producing the precise high-purity material demanded by the reactor. On February 19, 1960, ten fuel elements for the first load of CIRUS at Trombay were fabricated. CIRUS achieved criticality at AEET on July 10, 1960. The reactor is designed to achieve low-burn-up on the natural uranium fuel by quicker refueling sequences. The reactor is capable of producing approximately 9 kilograms of plutonium in a year.

13.4 SPENT FUEL REPROCESSING

In July 1958, Prime Minister Nehru authorized a project to build a spent fuel reprocessing facility with a capacity of processing 20 tonnes of spent fuel a year. The capacity was designed to match the production capacity of CIRUS. The plant was based on the U.S. developed PUREX process and was commissioned in mid-1964. As a nation, India has always placed a premium on self-sufficiency. Due to its vast domestic resources of thorium but limited supplies of uranium, from the start of its nuclear program, India has always placed strong emphasis on the development of breeder reactor fuel cycles. This provided a peaceful rationale for developing a plutonium separation capability.

13.5 INDIA'S NUCLEAR FACILITIES

India has a large and diverse nuclear power program which essentially includes all components of a closed nuclear fuel cycle. These facilities stand as an epitome of possibilities that were realized starting with international technological support and limited domestic resources. India had been extremely forthcoming with details of almost all the aspects of the nuclear program with the exception of plutonium production.

13.5.1 Power Reactors

The first nuclear power project of India started with General Electric constructing and commissioning two Boiling Water Reactor (BWR) power plants at Tarapur (near Mumbai) in 1969. Soon India realized the difficulty in acquiring enriched uranium for these reactor types and believed that BWR's would ensure lifetime dependence on the U.S. for fuel needs.

Even before India's first nuclear power plant at Tarapur, Homi Bhabha and his team were suggesting the three-stage-program and looked into the potential of CANDU-type reactors. This carried the burden of acquiring heavy water for moderation and as a coolant but using indigenous natural uranium fuel. By the year 1998, there existed 12 nuclear power plants in India. Among the plants under safeguards, two were RAPS-1 & 2 and the Tarapur BWR's, which were under safeguards in reciprocity for fuel provision from a sequence of international fuel sources. The technology to build 220MWe rated heavy water reactor units was mastered and plans were laid to further the progress of nuclear power plants. Building of nuclear power reactors continued with the addition of six more plants including two of 540 MWe ratings.

India currently possesses 18 operational PHWRs and 2 operational BWRs generating 4780 Mwe. Table 13.1 lists all the existing Indian nuclear power plants. India, being alienated from the advanced nuclear technology that it badly needed, was faced with technological hurdles following the 1974 nuclear test. Over a period of time, India obtained self sufficiency in its indigenous PHWR nuclear power plant technology. Capacity factors of the order of 80% were achieved in 2003. Four PHWRs of 700MWe along with 2VVERs (Russian PWR) of 1000MWe are under construction.

Table 13.1: India's Power Reactors

PLANT/UNITS/LOCATION	TYPE/CAPACITY
TARAPUR(TAPS 1,2), Maharashtra	BWR/160 * 2
RAJASTAN(RAPS1 to 6), Rajastan	PHWR/220 * 7
MADRAS(MAPS 1,2) Tamilnadu	PHWR/220 * 2
NARORA (NPS1,2) Uttar Pradesh	PHWR/220 * 2
KAIGA(KGS 1 to 4) Karnataka	PHWR/220 * 4
KAKRAPARA(KAPS 1,2) Gujarat	PHWR/220 * 2
TARAPUR(TAPS 3,4) Maharashtra	PHWR/540 * 2

India has an aggressive fast breeder reactor program. Technology demonstration was performed with the Fast Breeder Test Reactor (FBTR). While the design was a combination of

RAPSODIE and PHENIX reactors in France, all the components were manufactured within the country. It was commissioned in 1985. Based on the operating experience from FBTR, design of a 500 MWe demonstration reactor was started in 1990s. The initial design was for a pool type reactor with four secondary loops, but as the design progressed, it was desired to reduce the number of components with a view to minimize the manufacturing and construction time. Towards this a revised design with two secondary loops was carried out. Construction was started on 500 MWe Prototype Fast Breeder Reactors (PFBR) in 2006. The reactor is expected to be operational in 2013. There is a proposal to construct 4 more PFBRs by 2020, on the basis of twin reactors at a site.

13.5.2 Research Reactors

India has a number of research reactors. Construction of research reactor facilities began with the APSARA and CIRUS reactors. To advance research on the development of the second stage power reactor systems, the Fast Breeder Test Reactor(FBTR) was built based on the French Rapsodie design. This 40 MWth fast reactor was made operational with a mix of plutonium and uranium carbide as fuel. Technological demonstration of U-233 based reactor, which is crucial to the third stage of the three-stage program, was implemented with the commissioning and operation of the 30 kWth KAMINI reactor. Table 13.2 lists all the research reactors, including their capacity, type, date of commencement into service and their function.

Table 13.2: India's Research Reactors

Name/LOCATION	REACTOR POWER	START YEAR
CIRUS/TROMBAY	40 MWt, HEAVY WATER REACTOR	1960
DHRUVA/TROMBAY	100 MWt HEAVY WATER REACTOR	1985
APSARA/TROMBAY	1 MWt LIGHT WATER REACTOR	1956
PURNIMA 1/TROMBAY	CRITICAL ASSEMBLY 1W	1972
PURNIMA 2/TROMBAY	LIGHT WATER REACTOR 10W	1984
PURNIMA 3/TROMBAY	LIGHT WATER REACTOR 1W	1990
ZERLINA/TROMBAY	HEAVY WATER REACTOR, 100W	1961
FBTR/KALPAKKAM	40 MWt FAST BREEDER REACTOR	1985
KAMINI/KALPAKKAM	30 KWt LIGHT WATER REACTOR	1996

13.5.3 Uranium Enrichment Facilities

India's large scale uranium enrichment endeavor started in the early 1990's. The purpose of uranium enrichment in India is for nuclear submarine reactor applications and to provide fuel for the fast breeder reactors. The centrifuge process has been used in the Rare materials plant in Mysore. Research is in progress on other enrichment techniques like ultracentrifuge and laser enrichment.

13.5.4 Heavy Water Production

Table 13.3 lists the heavy water production units. There are, in total, eight heavy-water production plants with a capacity of more than 650 tonnes/yr, adequate both to support current domestic requirements and export sales. Six of these plants operate on ammonia exchange processes and two on the hydrogen sulfide process.

Table 13.3: Heavy Water Plants

PLANT	PROCESS	YEAR OF COMMENCEMENT
BARODA	AMMONIA EXCHANGE	1980
HAZIRA	AMMONIA EXCHANGE	1991
NANGAL	AMMONIA EXCHANGE	1962
TALCHER	AMMONIA EXCHANGE	1985
THAL-VASHISHT	AMMONIA EXCHANGE	1985
TUTICORIN	AMMONIA EXCHANGE	1978
KOTA	HYDROGEN SULPHIDE	1981
MANUGURU	HYDROGEN SULPHIDE	1991

13.5.5 Fuel Fabrication

The Nuclear Fuel Complex (NFC) at Hyderabad is the only large scale CANDU fuel fabrication facility in India. The NFC has an annual handling capacity of 250 tonnes of UF6 and the estimated capacity as input to the plant is 216 tonnes of UF6, after losses. In 2006, the NFC raised its handling capacity from 250 to 600 tonnes of UF6 per year.

The higher capacity though can only cater to the need of 14 PHWRs operating at 90% capacity factor. Any further addition of PHWR's would require additional rise in capacity. In addition to the mining of Uranium in new mines at Nalagonda in Andhra Pradesh and Meghalaya, there is a move to import natural uranium to tide over the shortage.

13.5.6 Reprocessing Facilities

These facilities are necessary for plutonium production for the second stage of India's three-stage program. Facilities are located at Trombay, Tarapur and Kalpakkam. Over the years, the research and development work carried out at Trombay has led to the setting up of improved designs. There is also a reprocessing development programme at Kalpakkam, which has successfully reprocessed the mixed PU-U carbide fuel. This facility was also used to reprocess the irradiated thorium fuel rods to extract U233. The facility at Kalpakkam would meet the fuel needs of the PFBR reactor under construction at Kalpakkam and the proposed fast breeders. With the proposal to construct two more 500 MWe Fast Reactors at Kalpakkam, construction of a co located Fast Reactor Fuel Cycle Facility (FRFCF) has begun. This will have complete reprocessing and fuel fabrication facilities for fast reactor fuels.

13.5.7 Uranium Mining, Milling

Uranium mining was initiated with governmental backing. Beginning with Jaduguda, six to seven different mining locations were discovered over a period of time. The Jaduguda mine, located in the Singhbum East District of Jharkand (formerly the southern portion of undivided Bihar state), began operations in 1967. It was long the primary source of Indian uranium.

13.6 FUTURE INDIAN STAGE 3 PROGRAMME

The objective of stage 3 is to achieve a sustainable nuclear fuel cycle by developing thorium–U-233 based systems that utilize India's vast thorium reserves to provide long-term energy security with nuclear power. New systems are being engineered to optimize the use of plutonium produced in stage 2 fast breeder reactors:

(a) to maximize the conversion of thorium to U-233,
(b) to extract power in-situ from the thorium fuel, and,
(c) to recycle the bred U-233 in additional reactors.

In addition, systems based on the thorium fuel cycle offer both neutronic and non-proliferation advantages over plutonium fuel cycles. These stage 3 concepts are to be implemented in parallel with the continuing development and deployment of stage 1 and stage 2 reactors and fuel cycle operations.

India has three innovative nuclear concepts in the design and development phases. The primary system for implementing the thorium utilization strategy is an advanced thermal reactor that draws on the proven PHWR pressure tube and heavy water technologies to satisfy each of the three design objectives stated above. A second, non-reactor, technology utilizes an accelerator and a subcritical assembly not only for the efficient conversion of thorium and possible generation of power but also for the incineration of long-lived actinides and fission products obtained from spent fuel reprocessing. The third system is a compact modular reactor suitable either for the production of electrical energy in remote areas or for the generation of process heat for the conversion of fossil fuels. Each of these concepts is discussed in detail below.

If these stage 3 systems achieved the goals of good economic performance and high levels of passive safety, India could begin large scale commercial deployment for electrical generation as well as for non-electrical applications, such as the desalination of sea water and the generation of portable, non-fossil fuels.

India's early stage 3 initiatives have encompassed small scale activities in all relevant technologies for the entire thorium fuel cycle. This includes the successful development and operation of KAMINI, the Kalpakkam Mini Reactor, the only U-233 fuelled reactor in the world currently in operation. Constructed at the IGCAR with fuel that was bred, processed, and fabricated indigenously, this reactor achieved criticality in 1996 and began full power operation at 30 kWt the following year. In addition to its mission as a U-233 fueled test reactor, it also functions as a neutron source for radiography and activation analysis. These modest but important indigenous research activities have laid the foundation for the development of the thorium fuel cycle and signaled India's entrance into the third stage of its nuclear power program.

13.6.1 Advanced Heavy Water Reactor (AHWR)

The primary reactor system envisioned for stage 3, the Advanced Heavy Water Reactor (AHWR), contains both evolutionary and revolutionary design concepts. The AHWR is a 920 MWt / 300 MWe heavy water moderated but light water cooled reactor that uses the well-proven pressure tube technology of the PHWR. The low-pressure reactor vessel, the "calandria," and the concentric calandria and pressure tubes are similar in design to those in the PHWR, except the calandria is oriented vertically, rather than horizontally. Vertical pressure tubes allow the removal of core heat through natural circulation of the boiling light water coolant, avoiding the need for primary coolant pumps and, hence, adding a measure of operational reliability and passive safety. The schematic flow diagram for the AHWR power plant is presented in Fig. 13.2.

In addition to power generation, the AHWR is also intended to desalinate sea water at the rate of 500 cubic meters (132,000 gal) per day. If desired, desalination capacity can be increased, with each 1000 cubic meters (264,000 gal) per day reducing the gross electrical output an estimated 0.95 MWe.

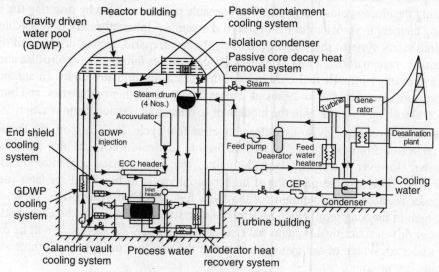

Fig. 13.2: Schematic Flow Sheet of AHWR Power Plant

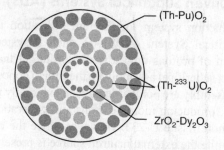

Fig. 13.3: Cross Section of AHWR fuel Cluster

The AHWR core, which is housed in the low-pressure calandria, contains 452 pressure tubes, each loaded with an identical fuel cluster. This cluster, as shown in Fig.13.3 consists of three concentric rings of fuel pins around a central rod assembly. The twenty-four pins in the outer ring are loaded with a mixture of thorium and plutonium oxides (ThO_2-PuO_2); in order to obtain "favorable minimum critical heat flux ratios," the plutonium content is 4.0% in the lower half of the active fuel and 2.5% in the upper half. The other two rings of fuel pins contain a mixture of thorium and U-233 oxides (ThO_2–U-233O_2) with a U-233 content of 3.0 % in the twelve pins of the inner ring and 3.75% in the eighteen pins of the middle ring. The central rod assembly contains twelve pins of dysprosium oxide in a zirconium dioxide matrix (DyO_2-ZrO_2) and a central channel for water from the Emergency Core Cooling System (ECCS). The light water coolant flows through the cluster in the spaces among the three concentric rings of fuel pins, but outside the central rod assembly. Maximum breeding of U-233 is needed to produce sufficient fissile material for recycle in AHWRs in order to attain the self-sufficiency characteristic required from a stage 3 reactor design. Power extracted in-situ is important for minimizing the initial inventory and consumption of the plutonium, while maintaining an average burnup of 24,000 MWd/t at discharge.

Inherent safety means, most importantly, ensuring a negative void coefficient of reactivity under both operating and accident conditions. This is achieved by hardening the neutron spectrum and inserting a burnable poison. The harder spectrum is obtained by decreasing the relative quantity of the heavy water moderator in the vicinity of the fuel cluster, that is, by positioning the clusters closer together. The burnable poison is added by inserting the neutron absorbing element dysprosium in the central rod assembly. Inherent safety is also achieved by engineered safety systems that actuate passively. In one scenario, depressurization of the main heat transport system during a loss of coolant accident causes failure of a rupture disc and floods the central channel in each fuel cluster from the ECCS accumulator tanks. In another, high steam pressure caused by loss of heat sink causes failure of another rupture disc and introduces a neutron absorbing "poison" into the moderator to quench the fission chain reaction.

The AHWR concept envisions a closed nuclear fuel cycle. Initially, both the thorium and U-233 recovered from the spent fuel will be used to fabricate fresh fuel pins for the AHWR fuel cluster. However, the reprocessed plutonium, with its increased concentration of higher plutonium isotopes and the presence of higher actinides, is to be stored for later fueling of fast breeder reactors; the plutonium for the fresh fuel will be obtained from recycled spent PHWR fuel. In the long term, when transmutation systems based on fast breeder reactors and accelerator driven subcritical systems have sufficient capacity, the fuel cycle will be extended so as to take advantage of the synergies between the various concepts in all three stages of India's nuclear program.

13.6.2 Accelerator Driven Subcritical Systems (ADS)

A second power production system for thorium utilization is being investigated – an Accelerator Driven Subcritical System (ADS). In such a system, a high current proton accelerator delivers a beam of protons onto a spallation target that functions as an external source of neutrons to drive a subcritical assembly. In the Indian concept, fissile U-233 and fertile Th-232 in the subcritical assembly are bombarded by the externally produced neutrons, causing either fission in the U-233, producing more neutrons and generating heat, or neutron capture by the Th-232, breeding more fissile U-233. However, since the configuration is subcritical, these processes continue while the external neutron source is present but decay away when the accelerator is turned off.

The ADS could be a technology for providing not only a self sufficient power production system but also a method to help alleviate problems associated with waste disposal. A properly designed ADS system will be able, in principle, to produce more fissile material than is consumed, also offering the promise of being able to reduce the doubling time for fissile material production, even with thorium. In addition, the fissions generate a sufficient heat to produce several times more electricity than is needed in the operation of the facility. An ADS system also affords the possibility of reducing the quantity of high-level nuclear waste that requires long-term geologic storage by transmuting long-lived actinides and fission products separated during the recycling operations into shorter-lived isotopes; this technology is being investigated, using both accelerators and fast reactors, in research laboratories in many nations.

India has outlined a phased approach to develop an ADS system, including critical facility measurements to validate relevant neutronic data, design and construct of cyclotron and linear accelerator facilities, and development of a molten, heavy metal spallation target. Physics design of the Low Energy High Intensity Proton Accelerator (LEHIPA) has been completed and simulation studies are being conducted on the main magnet of the 10 MeV cyclotron at the Variable Energy Cyclotron Center. Development of the spallation source is also in-progress.

13.6.3 Compact High Temperature Reactor (CHTR)

India is also developing a small modular reactor concept, the Compact High Temperature Reactor (CHTR), as an integral component of the stage 3 objective of utilizing its thorium resources to satisfy various energy needs. In addition, its development is serving as a demonstration of technologies relevant for next generation high temperature reactor systems.

The CHTR is intended to be able to produce either electrical power or process heat. It's compact, modular design makes it suitable for supplying electrical power to remote, difficult to access areas not connected to the nation's power grid. Its high operating temperature provides process heat suitable for the production of alternative transportation fuels such as hydrogen and the refinement of low–grade coal and oil to recover liquid fuels. Both forms of the CHTR's energy output can be used concurrently in the cogeneration of electricity and potable water.

The CHTR reactor is fuelled mainly with mixed uranium-thorium carbide having a U-233 content of 33.75%, moderated with beryllium oxide, and cooled with metallic lead-bismuth eutectic alloy. The reactor is able to operate without on-site refueling for an estimated fifteen years of continuous duty at full power. The current design operates at 1000°C but generates only 100 kWt in this, the design prototype; larger units are to be explored. The CHTR has passive systems for reactor control, shutdown, and heat removal under normal and postulated accident conditions. In addition, there are many inherent safety features: strong negative Doppler coefficient for the fuel, high thermal inertia and low power density of the core, large temperature margin on fuel design, negative temperature coefficient for the moderator, etc.

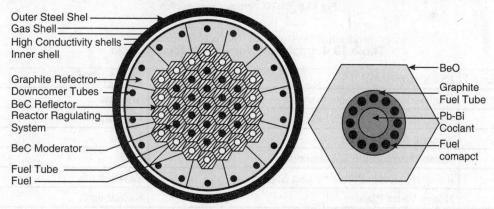

Fig. 13.4: CHTR core Plan and single fuel assembly

The CHTR core, shown in Fig.13.4 consists of three rings (nineteen elements) of hexagonal ("prismatic") beryllium oxide (BeO) "blocks," each containing a graphite "fuel tube" with twelve "fuel bores" around a central coolant channel. The fuel is made of TRISO coated fuel particles embedded in a graphite powder matrix loaded in cylindrical "fuel compacts" and inserted in the fuel bores. A ring of eighteen beryllium oxide blocks surround the core region serving as a reflector to scatter neutrons back into the core and housing a passive control / safety system responsive to outlet coolant temperature. In addition, graphite material in the outer ring contains channels to return the coolant to the bottom of the core.

This assemblage is contained in a shell of high-temperature resistant and liquid-metal-corrosion resistant material. A vertical cross-section diagram of the reactor identifying not only the various core elements but also the control / shutdown, coolant and structural components is presented in Fig. 13.5.

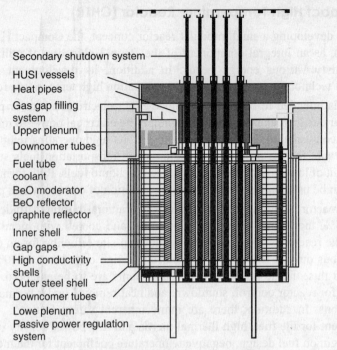

Fig.13.5: CHTR Vertical Cross section

Table 13.4: Important Nuclear Facilities in India

No	ESTABLISHMENTS	LOCATION
1	Nuclear Research Laboratory	Srinagar
2	High Altitude Research Laboratory	Gulmarg
3	Seismic Array Station	Kasan
4	Radio Pharmaceutical Laboratory	Delhi
5	Rajastan Atomic Power Project Station	Rawatbhatta
6	Heavy Water Plant	Rawatbhatta
7	Narora Atomic Power Station	Narora
8	Centre for Advanced Technology	Indore
9	Harish Chandra Research Institute	Allahabad
10	Gurushikar Observatory for Astrophysical Sciences	Mount Abu
11	Uranium Corporation of India Mine	Jaduguda/Bhatin/Turamdih/Narwapahar
12	Heavy Water Plant	Baroda
13	Institute of Plasma Research	Ahmedabad
14	Radioimmuno Assay Centre	Dibrugarh
15	Variable Energy Cyclotron Centre	Kolkata
16	Saha Institute of Nuclear Physics	Kolkata
17	Kakrapar Atomic Power Station	Kakrapar
18	Heavy Water Plant	Hazira

19	Tarapur Atomic Power Station	Tarapur
20	Waste Immobilisation Plant	Tarapur
21	Power Reactor Fuel Reprocessing Plant	Tarapur
22	Advanced Fuel Fabrication Plant	Tarapur
23	Heavy Water Plant	Talcher
24	Institute of Physics	Bhubaneshwar
25	Orissa Sand Complex	Chhatrapur
26	Heavy Water Plant	Managuru
27	Bhabha Atomic Research Centre	Mumbai
28	ISOMED Plant	Mumbai
29	Nuclear Fuel Complex	Hyderabad
30	Electronics Corporation of India	Hyderabad
31	Heavy Water Plant	Thal
32	Kaiga Atomic Power Station	Kaiga
33	Seismic Laboratory	Gauribidhanur
34	Rare Metals Plant	Mysore
35	Institute of Mathematical Sciences	Chennai
36	Madras Atomic Power Station	Kalpakkam
37	Indira Gandhi Centre for Atomic Research	Kalpakkam
38	Prototype Fast Breeder Reactor Project	Kalpakkam
39	Nuclear Desalination Plant	Kalpakkam
40	Fuel Reprocessing Plant	Kalpakkam
41	Rare Earths Plant	Udyogmandal
42	Heavy Water Plant	Tuticorin
43	Koodankulam Atomic Power station	Koodankulam

Heat is passively removed from the core under both normal and abnormal operating conditions. During normal operations, the lead-bismuth coolant flows by natural circulation from the lower to the upper plenum through the central channels in the fuel tubes, returning through the "downcomers" at the reactor periphery. A set of sodium heat pipes in the upper plenum then passively transfers this heat with a minimum temperature loss to heat-utilization vessels which provide the interface to systems for high temperature heat applications.

There are also three passive safety systems for heat removal in the event of an abnormal condition or accident. One is a siphon system to flood gas-filled gaps in the core with molten metal to facilitate heat conduction to the outside. Another is a set of six variable conductance heat pipes located in the upper plenum to transfer the core's heat to the atmosphere under the loss of load condition when the primary coolant circuit remains intact. The last is a system of twelve carbon composite variable conductance heat pipes located in the core that service the need when the primary coolant is lost. These systems, each of which is individually capable of removing the equivalent of 200 kWt, can act together or independently to limit the temperature increase in the core and coolant.

SUMMARY

India's nuclear energy self-sufficiency extended from uranium exploration and mining through fuel fabrication, heavy water production, reactor design and construction, to reprocessing and waste management (Table 13.4). However, India is outside the Nuclear Non-Proliferation Treaty and it has been for 34 years largely excluded from trade in nuclear plant or materials. Due to these trade bans and lack of indigenous uranium, India has uniquely been developing a nuclear fuel cycle to exploit its reserves of thorium. Now, foreign technology and fuel are expected to boost India's nuclear power plans considerably. All plants will have high indigenous engineering content. India has a vision of becoming a world leader in nuclear technology due to its expertise in fast reactors and thorium fuel cycle.

BIBLIOGRAPHY

1. S.K.Jain, Nuclear Power – An alternative, http://www.npcil.nic.in/
2. A. Kakodkar ,Evolving Indian Nuclear Programme: Rationale and Perspectives, Available at http://www.dae.nic.in/(2008)
3. Shapingthe Third Stage of Indian Nuclear Power Programme,http://www.dae.nic.in/
4. Milestones achieved by Department of Atomic Energy in India, http://www.dae.nic.in/
5. Advanced Heavy Water Reactor, www.dae.nic.in.
6. Radiation – Sources,Technologies and Applications for Societal Development, http://www.dae.nic.in/
11. R.K.Sinha, Anil Kakodkar, Design and Development of the AHWR-The Indian thorium fuelled innovative nuclear reactor, Nuclear Engineering and Design, 236(2006)683-700.
12. S.A.Bohra, P.D.Sharma, Construction management of Indian Pressurised Heavy water reactors, Nuclear Engineering and Design, 236(2006)836-851.
13. S.K. Agarwal,C.G. Karhadkar, A.K.Zope, Kanchhi Singh, DHRUVA: Main Design Features, operational Experience and utilization, Nuclear Engineering and Design, 236(2006)747-757.
14. V.K. Raina, R. Srivenkatesan, D.C. Khatri, D.K.Lahiri, Critical Facility for Lattice Physics Experiments for AHWR and 500 MWe PHWR, Nuclear Engineering and Design, 236(2006) 758-769.
15. S.K.agarwal, Ashok Chauhan, Alok Mishra, The VVERs at Kudankulam, Nuclear Engineering and Design, 236(2006)812-835.

ASSIGNMENTS

1. Explainin detail the three stage nuclear power programme envisaged by Dr Homi Bhabha.
2. What led to the choice of PHWR for India's power programme as compared to PWR/BWR.
3. What are different research reactors and what was their role.
4. Whatare the different types of reactors under operation in India and where are they located.
5. What is the role of Advanced heavy water reactor in the three stage programme.

Next Generation of Reactors

14.0 INTRODUCTION

In any technology, feedbacks obtained from the current ones and the knowledge base created are used to design better economic products. Without scientists, there are no technical advances. Without engineers, there are no products. One of the greatest challenges for a technology is to focus its R&D investments in areas with the greatest potential payoff. Such is also the case for the nuclear power industry. The evolution of the reactors has followed a progressively encouraging path in that safety has improved with every design. The reactors are grouped under different generations (Fig 14.1). This chapter summarizes the different nuclear power systems that are in operation and those under development and competing for attention and investment.

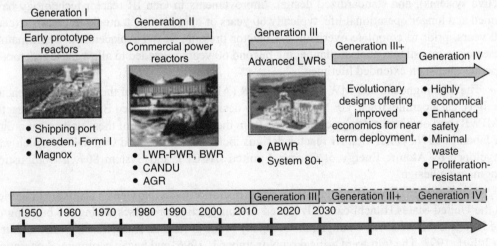

Fig. 14.1: Generations of Nuclear Reactors

14.1 GENERATION I

Gen I refers to the prototype and power reactors that launched civil nuclear power. This generation consists of early prototype reactors from the 1950s and 1960s, such as Shippingport (1957–1982) in Pennsylvania, Dresden-1 (1960–1978) in Illinois, and Calder

Hall-1 (1956–2003) in the United Kingdom. This kind of reactor typically ran at power levels that were "proof-of-concept." The only remaining commercial Gen I plant, the Wylfa Nuclear Power Station in Wales, was scheduled for closure in 2010. However, the UK Nuclear Decommissioning Authority announced in October 2010 that the Wylfa Nuclear Power Station will operate up to December 2012.

14.2 GENERATION II

Gen II refers to a class of commercial reactors designed to be economical and reliable. Designed for a typical operational lifetime of 40 years, prototypical Gen II reactors include pressurized water reactors (PWR), CANada Deuterium Uranium reactors (CANDU), boiling water reactors (BWR), advanced gas-cooled reactors (AGR), and Vodo-Vodyanoi Energetichesky Reactors (VVER).

Gen II systems began operation in the late 1960s and comprise the bulk of the world's 400+ commercial PWRs and BWRs. These reactors, typically referred to as light water reactors (LWRs), use traditional active safety features involving electrical or mechanical operations that are initiated automatically and, in many cases, can be initiated by the operators of the nuclear reactors. Some engineered systems still operate passively (for example, using pressure relief valves) and function without operator control or loss of auxiliary power. Most of the Gen II plants still in operation in the West were manufactured by one of three companies: Westinghouse, Framatome (now part of AREVA), and General Electric (GE).

14.3 GENERATION III

Gen III nuclear reactors are essentially Gen II reactors with evolutionary, state-of-the-art design improvements. These improvements are in the areas of fuel technology, thermal efficiency, modularized construction, safety systems (especially the use of passive rather than active systems), and standardized design. Improvements in Gen III reactor technology have aimed at a longer operational life, typically 60 years of operation, potentially to greatly exceed 60 years, prior to complete overhaul and reactor pressure vessel replacement. Confirmatory research to investigate nuclear plant aging beyond 60 years is needed to allow these reactors to operate over such extended lifetimes.

The Westinghouse 600 MW advanced PWR (AP-600) was one of the first Gen III reactor designs. On a parallel track, GE Nuclear Energy designed the Advanced Boiling Water Reactor (ABWR) and obtained a design certification from the NRC. The first of these units went online in Japan in 1996. Other Gen III reactor designs include the Enhanced CANDU 6, which was developed by Atomic Energy of Canada Limited (AECL); and System 80+, a Combustion Engineering design.

Only four Gen III reactors, all ABWRs, are in operation today. No Gen III plants are in service in the United States. Hitachi carefully honed its construction processes during the building of the Japanese ABWRs. For example, the company broke ground on Kashiwazaki-Kariwa Unit 7 on July 1, 1993. The unit went critical on November 1, 1996, and began commercial operation on July 2, 1997—four years and a day after the first shovel of dirt was turned.

14.3.1 ABWR

The ABWR represents a more evolutionary route for the BWR family, with numerous changes and improvements to the standard BWR design.

Major areas of improvement include:
- The addition of reactor internal pumps (RIPs) to the bottom of the RPV (reactor Pressure vessel) -10 in total- which achieve improved performance while eliminating large-diameter and complex piping structures at the bottom of the RPV (e.g. the recirculation loop found in earlier BWR models). Only the RIP motor is located outside of the RPV in the ABWR.
- The control rod adjustment capabilities have been supplemented with the addition of the electro-hydraulic Fine Motion Control Rod Drive (FMCRD), allowing for fine position adjustment, while not losing the reliability or redundancy of traditional hydraulic systems which are designed to accomplish shutdown in 2.80 seconds from receipt of an initiating signal.
- A fully digital Reactor Protection System (with redundant digital backups as well as redundant manual backups) ensures a high level of reliability and simplification for safety condition detection and response. Standard BWR 2 out of 4 rapid shutdown logic ensures that spurious rapid shutdowns are not triggered by single instrument failures.
- Fully digital reactor controls (with redundant digital backup and redundant manual backups) allow the control room to easily and rapidly control plant operations and processes. Separate, redundant critical and non-critical digital multiplexing buses allow for reliability and diversity of instrumentation and control.
- In particular, the reactor can both "fly on autopilot" and also "take off and land on autopilot" or go critical and ascend to power using automatic systems only and do a standard shutdown using automatic systems only. Of course, human operators remain essential to reactor control, but much of the busy-work of bringing the reactor to power and descending from power can be automated at operator discretion.
- The emergency core cooling systems (ECCS) has been improved in many areas, providing a very high level of defence-in-depth against accidents, contingencies, and incidents.
  - The overall system has been divided up into 3 divisions; each division is capable- by itself-of reacting to the maximally contingent Limiting Fault/Design Basis Accident (DBA) and terminating the accident prior to core uncovering, even in the event of loss of offsite power and loss of feedwater. Previous BWRs had 2 divisions, and uncovering (but no core damage) was predicted to occur for a short time in the event of a severe accident, prior to ECCS response.
  - Eighteen SORVs (safety overpressure relief valves), ten of which are part of the ADS (automatic depressurization system), ensure that RPV overpressure events are quickly mitigated, and that if necessary, that the reactor can be depressurized rapidly to a level where low pressure core flooder (LPCF, the high-capacity mode of the residual heat removal system, which replaces the LPCI and LPCS in previous BWR models) can be used.
  - Further, LPCF can inject against much higher RPV pressures, providing an increased level of safety in the event of intermediate-sized breaks, which could be small enough to result in slow natural depressurization but could be large enough to result in high pressure core spray/coolant injection systems' capacities for response being overwhelmed by the size of the break.

- Though the Class 1E (life safety critical) power bus is still powered by 3 highly-reliable emergency diesel generators that are safety rated, an additional Plant Investment Protection power bus using a combustion gas turbine is located on-site to generate electricity to provide defence in depth against station blackout contingencies as well as to power important but non-safety critical systems in the event of a loss of offsite power, as well as to start the plant in the event grid black start is needed. Additional diesel feedwater pumps may be tied into the plant's service water system too, to enhance cooling capabilities.
- The containment has been significantly improved over old BWR types. Like the old types, it is of the pressure suppression type, designed to handle evolved steam in the event of a transient, incident, or accident by routing the steam using pipes that go into a pool of water, called the wetwell (or torus), the low temperature of which will condense the steam back into liquid water. This will keep pressure low. Notably, the typical ABWR containment has numerous hardened layers between the interior of the primary containment and the outer shield wall, and is cubical in shape. One major enhancement is that the reactor has a standard safe shutdown earthquake acceleration of .2 G (slightly less than 2 m/s^2); further, it is designed to withstand a tornado of Old Fujita Scale 6, with > 320 mph wind). Seismic Hardening is possible in earthquake-prone areas and has been done at the Lungmen facility in Taiwan which has been hardened up .3 G (slightly less than 3 m/s^2) in any direction.
- The ABWR is designed for a lifetime of at least 60 years, though operation beyond that 60 year point will certainly be possible unless safety limits within the expensive to replace reactor pressure vessel is reached. The comparatively simple design of the ABWR also means that no expensive steam generators need to be replaced, either, decreasing total cost of operation.

The ABWR is fully automated in response to a Loss of Coolant Accident (LOCA), and operator action is not required for 3 days. After 3 days the operators must replenish ECCS water supplies. These and other improvements make the plant significantly safer than previous reactors.

14.4 GENERATION III+

Gen III+ reactor designs are an evolutionary development of Gen III reactors, offering significant improvements in safety over Gen III reactor designs certified by the NRC in the 1990s. Examples of Gen III+ designs include:

- VVER-1200/392M Reactor of the AES-2006 type
- Advanced CANDU Reactor (ACR-1000)
- AP1000: based on the AP600, with increased power output
- European Pressurized Reactor (EPR): evolutionary descendant of the Framatome N4 and Siemens Power Generation Division KONVOI reactors
- Economic Simplified Boiling Water Reactor (ESBWR): based on the ABWR
- APR-1400: an advanced PWR design evolved from the U.S. System 80+, originally known as the Korean Next Generation Reactor (KNGR)
- EU-ABWR: based on the ABWR, with increased power output and compliance with EU safety standards
- Advanced PWR (APWR): designed by Mitsubishi Heavy Industries (MHI)

- ATMEA I: a 1,000–1,160 MW PWR, the result of collaboration between MHI and AREVA.

Manufacturers began development of Gen III+ systems in the 1990s by building on the operating experience of the American, Japanese, and Western European LWR fleets. Perhaps the most significant improvement of Gen III+ systems over second-generation designs is the incorporation in some designs of passive safety features that do not require active controls or operator intervention but instead rely on gravity or natural convection to mitigate the impact of abnormal events. The inclusion of passive safety features, among other improvements, may help expedite the reactor certification review process and thus shorten construction schedules. These reactors, once on line, are expected to achieve higher fuel burnup than their evolutionary predecessors (thus reducing fuel consumption and waste production). More than two dozen Gen III+ reactors based on five technologies are planned for the United States.

14.4.1 EPR

Areva NP (formerly Framatome ANP) has developed a large (4590 MWt, typically 1750 MWe gross and 1630 MWe net) European pressurised water reactor (EPR) which was confirmed in mid 1995 as the new standard design for France and received French design approval in 2004. It is a 4-loop design derived from the German Konvoi types with features from the French N4, and is expected to provide power about 10% cheaper than the N4. It will operate flexibly to follow loads, have fuel burn-up of 65 GWd/t and a high thermal efficiency, of 37%, and net efficiency of 36%. It is capable of using a full core load of MOX. Availability is expected to be 92% over a 60-year service life.

It has double containment with four separate, redundant active safety systems, and boasts a core catcher under the pressure vessel. The safety systems are physically separated through four ancillary buildings on the same concrete raft, and two of them are aircraft crash protected. The primary diesel generators have fuel for 72 hours, the secondary back-up ones for 24 hours, and tertiary battery back-up lasts 12 hours. It is designed to withstand seismic ground acceleration of 600 G without safety impairment.

The first EPR unit is being built at Olkiluoto in Finland, the second at Flamanville in France, the third European one will be at Penly in France, and two further units are under construction at Taishan in China. A similar unit is planned to be built at Jaitapur, Maharashtra in India.

A US version, the US-EPR quoted as 1710 MWe gross and about 1580 MWe net, was submitted for US design certification in December 2007, and this is expected to be granted early 2013. The first unit (with 80% US content) was expected to be grid connected by 2020. It is now known as the Evolutionary PWR (EPR).

14.5 GENERATION IV CONCEPTS

Nuclear scientists have left implementation of the Gen III+ designs in steel and concrete to the engineers and moved on to developing the "generation after next" nuclear alternatives—commonly called Gen IV. Conceptually, Gen IV reactors have all of the features of Gen III+ units plus the ability to support hydrogen production, thermal energy off-taking, and perhaps even water desalination. In addition, these designs include advanced actinide management. An actinide is an element with an atomic number between 89 (actinium) and 103 (lawrencium); the term is usually applied to elements heavier than uranium, which are also called transuranics. Actinides are radioactive, typically have long half-lives, and constitute a significant portion of the spent fuel wastes from LWRs.

The table 14.1 summarizes the characteristics and operating parameters of six Gen IV reactor system alternatives, including the molten salt reactor. Each of the technology concepts has been prioritized to reflect its technology development status and its potential to meet the program's and national goals. In general, Gen IV systems include full actinide recycling and on-site fuel cycle facilities based on either advanced aqueous, pyro-metallurgical, or other dry processing options. On-site reprocessing minimizes the transportation of nuclear materials, reducing the chances of proliferation. Following are brief description of the development status of the six Gen IV reactor system alternatives.

Table 14.1: GEN IV CONCEPTS

Reactor	Neutron Spectrum	Coolant	Temperature°C	Pressure	Fuel	Fuel Cycle	Power MW
Gas Cooled Reactor	Fast	Helium	850	high	Mixed oxide	Closed	280
Lead Cooled Reactor	Fast	Pb-Bi	550-800	low	-do-	closed	50-1200
Molten Salt Reactor	Epithermal	Fluoride Salts	700-800	low	UF in Salt	Closed	1000
Sodium Cooled Reactor	Fast	Sodium	550	low	Mixed oxide	Closed	150-1500
Super Critical Water Cooled Reactor	Thermal /Fast	Water	510-550	Very High	UO2	Open/ Closed	1500
Very High Temperature Reactor	Thermal	Helium	1000	high	UO2-Prism/ Pebbles	Open	250

14.5.1 Gas-cooled fast reactor (GFR)

The GFR (Figure 14.2) is primarily designed for electricity production and actinide management, but it may be able to support hydrogen production as well. The reference GFR system features a fast neutron spectrum, a Brayton-cycle helium-cooled reactor, a closed fuel cycle for actinide reprocessing, and a plant efficiency of 48%. In November 2006, the GFR System Arrangement was signed by the European Atomic Energy Community (Euratom), France, Japan, and Switzerland.

The several forms of fuel (ceramics, fuel particles, and ceramic-clad elements) being considered for the GFR has one thing in common: They will allow the reactor to operate at very high temperatures yet ensure excellent containment of fission products. Core configurations will be either pin-or plate-based fuel assemblies or prismatic blocks. Performance enhancement possibilities still being researched include the use of materials with superior resistance to fast neutron fluence (flux integrated over time) at very high temperatures, and the development of a helium-cooled turbine capable of super-efficient electricity production. Target values of some key parameters, such as power density and fuel burn-up, are sufficient for reasonable performance of a first-generation technology.

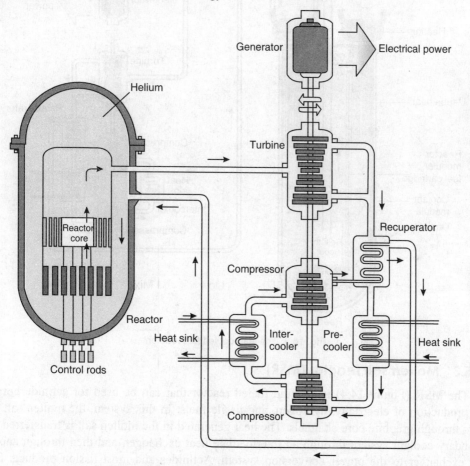

Fig. 14.2: The gas-cooled fast reactor

14.5.2 Lead-cooled fast reactor (LFR)

The LFR (Figure 14.3) is a fast neutron spectrum reactor designed for electricity and hydrogen production as well as actinide management. Three key technical aspects of the LFR are its use of lead for cooling, a long cartridge-core life (15 to 20 years), and its modularity and small size (potentially suiting it for deployment on small grids or at remote locations).

The LFR envisioned by Generation IV program would be based on the small secure transportable autonomous reactor (SSTAR) concept. The main mission of SSTAR development is to provide incremental energy generation to match the needs of developing nations and remote communities lacking a grid connection. Lead cooled reactor technologies have already been successfully demonstrated in Russia, with 40 years of experience in lead-bismuth cooling of the reactors powering its Alfa-class submarines.

The LFR relies on natural circulation to cool the reactor. An indirect gas Brayton cycle is also incorporated is produce maximum electrical power.

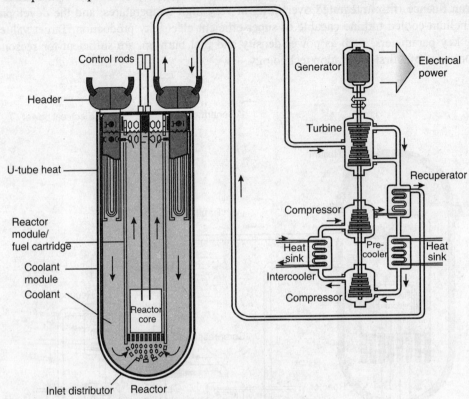

Fig. 14.3: The lead-cooled fast reactor.

14.5.3 Molten salt reactor (MSR)

The MSR (Figure 14.4) is a liquid-fueled reactor that can be used for actinide burning and production of electricity, hydrogen, and fissile fuels. In this system, the molten salt fuel flows through graphite core channels. The heat generated in the molten salt is transferred to a secondary coolant system through an intermediate heat exchanger, and then through another heat exchanger to the power conversion system. Actinides and most fission products form fluorides in the liquid coolant. The homogenous liquid fuel allows for the addition of actinide feeds without requiring fuel fabrication. The power conversion module could be either an gas turbine cycle or a standard steam power cycle.

During the 1960s, the U.S. developed a molten salt breeder reactor as the primary back-up option for a conventional fast breeder reactor. Recent work has focused on lithium and beryllium fluoride coolants with dissolved thorium and U-233 fuel. The DOE plans to continue its cooperative work with Euratom MSR programs in the future.

14.5.4 Sodium-cooled fast reactor (SFR)

The primary development goals of the SFR (Figure 14.5) program are actinide management, reduction of waste products, and more-efficient uranium consumption. Future, lower-cost designs are expected to not only produce electricity but also supply thermal energy, produce hydrogen, and possibly enable desalination as well. The SFR's fast neutron spectrum could

Next Generation of Reactors

make the use of available fissile and fertile materials, including depleted uranium, much more efficient than it is in today's LWRs. In addition, the SFR system may not require as much design research as other Generation IV systems.

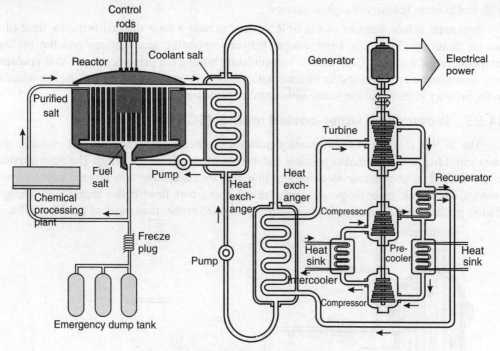

Fig. 14.4: The molten salt reactor

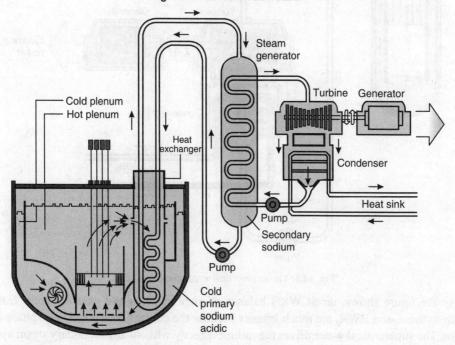

Fig. 14.5: The sodium-cooled fast reactor.

A Gen IV technical readiness and operating experience comparison of the GFR, LFR, and SFR systems led to the selection of the SFR as the primary fast-reactor Gen IV candidate for near-term deployment. The decision was based on more than 300 reactor-years' experience with fast neutron reactors in eight countries.

Important safety features of the SFR system include a long thermal response time (the reactor heats up slowly), a large margin between operating temperatures and the boiling temperatures of coolants (less chance of accidental boiling), a primary system that operates near atmospheric pressure, and an intermediate sodium system between the radioactive sodium in the primary system and the water and steam in the power plant.

14.5.5 Supercritical water-cooled reactor (SCWR)

The SCWR (Figure 14.6) promises significant economic advantages for two reasons: the plant simplification that it makes possible and its increased thermal efficiency. The main mission of the SCWR is to generate electricity at low cost by combining two proven technologies: conventional LWR technology and supercritical fossil fuel–fired boiler technology. Design studies predict plant thermal efficiencies about one-third higher than those of today's LWRs.

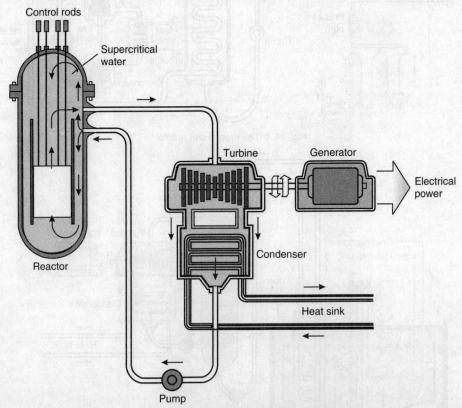

Fig. 14.6: The supercritical water-cooled reactor

As the figure shows, an SCWR's balance-of-plant systems and passive safety features, similar to those of a BWR, are much simpler because the coolant does not change phase in the reactor. The supercritical water drives the turbine directly without any secondary steam system. An international effort, with Japan in the lead, aims to resolve the most pressing materials and system design uncertainties needed to demonstrate the technical viability of the SCWR.

14.5.6 Very high temperature reactor (VHTR)

The main mission of the VHTR (Figure 14.7) is to produce both electricity and hydrogen. The reference system consists of a helium-cooled, graphite-moderated, thermal neutron reactor. Electricity and hydrogen are produced using an indirect cycle in which intermediate heat exchangers supply a hydrogen production demonstration facility and a gas turbine generator. Process heat also could be provided for applications such as coal gasification and cogeneration.

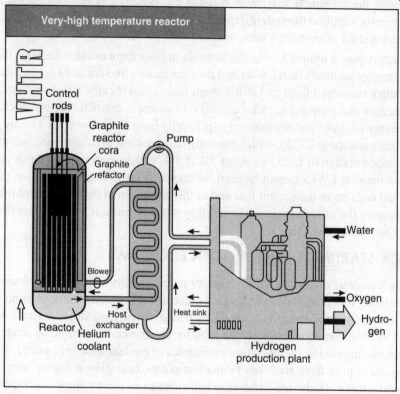

Fig. 14.7: The very high temperature reactor

The VHTR gets high economic marks for its high hydrogen production efficiency and high safety and reliability grades due to the inherent safety features of the fuel and reactor. It also gets good ratings for proliferation resistance and physical protection, and a neutral rating for sustainability because of its open or once-through fuel cycle. Although the VHTR requires R&D advances in fuel performance and high-temperature materials, it should benefit from earlier GFR, GT-MHR, and PBMR advancements.

14.6 ACTINIDE MANAGEMENT

Actinide management, common to all the Gen IV alternatives, would reduce the volume of nuclear waste in the mid-term and provide assurance of nuclear fuel availability in the long term. This mission overlaps a national responsibility addressed in the Nuclear Waste Policy Act, namely, the disposition of spent nuclear fuel and high-level waste. The mid-term (30 to 50 years) actinide management mission consists primarily of limiting or reversing the buildup of the inventory of spent nuclear fuel from current and near-term nuclear plants.

Actinides may be a waste product for an LWR, but they are fissionable in a fast reactor. As mentioned earlier, a transuranic is a very heavy element with a higher atomic number than uranium (92); it is formed artificially by neutron capture and possibly by subsequent beta decays. Extracting these long-lived radionuclides from spent fuel and irradiating them in a closed fuel cycle using fast reactors does more than generate electricity. It also transmutes the long-lived radio nuclides that would otherwise require isolation in a geologic repository such as Yucca Mountain into shorter-lived radio nuclides. Transmutation changes atoms of one element into those of another by neutron bombardment that causes neutron capture and/or fission. In the longer term, the actinide management mission can beneficially produce excess fissionable material, currently supplied through mining and the enrichment of natural uranium, for use in systems optimized for other energy missions.

Fast reactors play a unique role in the actinide management mission because they operate with higher-energy neutrons than LWRs and thus are more effective in fissioning the actinides and transuranics recovered from an LWR's spent fuel. Theoretically, a fast reactor can recycle all of the uranium and transuranic radio nuclides. In contrast, thermal reactors, such as LWRs, use lower-energy neutrons and extract energy primarily from fissile isotopes. The only naturally occurring fissile isotope is U-235, which has only 0.7% natural uranium; enrichment increases this natural concentration of U-235 to about 3% to 5%, which is enough to enable operation of an LWR. But because LWRs cannot be used for complete recycling, over 99% of the uranium initially mined ends up in their spent fuel and in the residue from the enrichment process. Fast reactors maximize the use of uranium because they support multiple fuel recycles that make all of the fuel's heat content usable.

14.7 KICK-STARTING THE HYDROGEN ECONOMY

Another feature of many of Gen IV reactors is their ability to produce hydrogen as a by-product. Realizing this potential could make the use of fuel cells for transportation and power generation more economic and environmentally benign while reducing America's dependence on imported oil. Sufficient quantities of hydrogen for commercial use would be produced during off-peak periods, improving the operating economics of nuclear base load plants. A long-term objective would require dedicated Gen IV nuclear plants, operating at higher temperatures, to produce hydrogen at a steady rate for storage and subsequent use by large (>1,000-MW) banks of fuel cells to address daily peak demand.

SUMMARY

This chapter has taken the reader through a tour of the four generations of nuclear reactors. Brief outline of generation III and IV reactors have been given to appreciate their more passive safety features as compared to the earlier generation of reactors. Such features are expected to result in better safety and make them amenable to better public acceptance. International work will be directed henceforth on these concepts.

BIBLIOGRAPHY

1. GEN IV Roadmap, http://gif.inel.gov/roadmap (2002)
2. J. Bouchard, "The Global View," GIF Symposium-Paris (France)-9-10 September, www.gen4.org/GIF/About/documents/ GIF ProceedingsWEB.pdf, (2009)

3. http://www.world-nuclear-news.org/NP-New_name_and_mission_for_GNEP-2106108.html, (2010)
4. "US GNEP Program Dead, DOE Confirms," Nuclear Engineering International, April 15, (2009)
5. http://www.iaea.org/INPRO/
6. Chapter 23 New Generation Reactors in Energy and Power Generation Handbook: Established and Emerging Technologies, Ed. K.R.Rao, ASME Digital Library(2011).
7. Kenneth D. Kok (ed.), Handbook of Nuclear Engineering, CRC Press, (2009).
8. Dan Gabriel Cacuci (ed.), Handbook of Nuclear Engineering, Springer (2010).
9. Steven B., Krivit, Jay H.Lehr, Thomas B. Kingery, Nuclear Energy Encyclopedia: Science, Technology, and Applications, John Wiley (2011).

ASSIGNMENTS

1. What are the different generations of Nuclear Reactors.
2. Whatare the features of GEN IV reactors that score over previous generation reactors.
3. Whatare the different reactor concepts chosen by GEN IV group of countries for further development.
4. Whatdo you mean by passive safety. Is a passive safety system more reliable than an active safety system.
5. Collect information on the French EPR reactors and present in a seminar.

Safety Approaches in Reactor Design

15.0 INTRODUCTION

Nuclear power plants design is carried out with three basic objectives: general nuclear safety objective, Radiation Protection objective and technical Safety objective. The general safety objective refers to protect the individuals, society and environment from harm by establishing and maintaining effective defenses in nuclear installations against radiological accidents. This objective is supported by two complementary objectives dealing with radiation protection and technical aspects. The Radiation protection objective is to ensure that in all operational states of a nuclear installation or due to any planned release of radioactive material from the installation, is kept below prescribed limits and as low as reasonably achievable and to ensure mitigation of the radiological consequences of any incidents. The technical safety objective necessitates the utility to take all practicable measures to prevent accidents in nuclear installations and to mitigate their consequences should they occur; to ensure with a high level of confidence that, for all postulated accidents considered in design of the installation, including those of very low probability, any radiological consequences would be minor and below prescribed limits and to ensure that the likelihood of accidents with serious radiological consequences is extremely low.

Safety objectives require that nuclear installations are designed and operated so as to keep all sources of radiation exposure under strict technical and administrative control. However, the radiation protection objective does not preclude limited exposure of the people or release of legally authorized quantities of radioactive materials to the environment from these installations. Such exposures and releases must be made in compliance with operational limits and radiation protection standards. This chapter deals with the safety approaches in the design and operation of nuclear facilities worldwide.

15.1 COMPREHENSIVE SAFETY ANALYSIS

To achieve these objectives in the design of a nuclear power plant comprehensive safety analysis are carried out to identify all sources of exposure and to evaluate the radiation doses that could be received by the public and by occupational workers, as well as the potential effects of radiation on the environment. The safety analysis examines:

- all planned normal operational modes of operation of the plant
- plant performance under anticipated operational occurrences

Safety Approaches in Reactor Design

- design basis events
- Event sequences that may lead to consequences beyond predicted levels

The design for safety ensures that plant states that could result in high radiation doses or radionuclide release are of very low probability and the plant states with significant probabilities of occurrence have only minor or no potential radiological consequences. It is essential to ensure in design that the need for external intervention is limited or even eliminated.

15.1.1 Defence in Depth

The safety objectives are achieved through the application of the defence in depth strategy. The strategy is to prevent occurrence of incidents and if they cannot be prevented, to limit their consequences and prevent evolution to more serious condition. Preventing the degradation of the plant status and performance will be most effective step in protection. Should preventive measures fail, control and mitigation with use of a well designed confinement system can provide the necessary protection of public and environment. If any failure were to occur, it should be detected and compensated by appropriate measures. Application of the concept of defence in depth in the plant design provides a series of levels of defence (Inherent features, equipment and procedures aimed at preventing accidents).

Table 15.1: LEVELS OF DEFENCE

LEVELS OF DEFENCE	OBJECTIVE	ESSENTIAL MEANS
LEVEL 1	Prevention of abnormal operation and failures	Conservative design and high quality in construction and operation
LEVEL 2	Control of abnormal operation and detection of failures	Control, limiting and protection systems and other surveillance features
LEVEL 3	Control of accidents within the design basis	Engineered safety features and accident procedures
LEVEL 4	Control of severe plant conditions including prevention of accident progression and mitigation of the consequences of severe accidents	Complementary measures and accident management
LEVEL 5	Mitigation of radiological consequences of significant releases of radioactive materials	Off-site emergency response

This strategy has proved to be effective in compensating for human and equipment failures, both potential and actual. There is no unique way to implement defence in depth (i.e. no unique technical solution to meet the safety objectives), since there are different designs, different safety requirements in different countries, different technical solutions. Nevertheless, the strategy represents the best general framework to achieve safety for any type of nuclear power plants.

Defence in depth is generally structured in five levels. Should one level fail, the subsequent level comes into play. Table 15.1, summarizes the objectives of each one of the five levels and the correspondent primary means of achieving them. The general objective of defence in depth is to ensure that a failure, whether equipment failure or human failure, at one level of defence, and even combinations of failures at more than one level of defence, would not propagate to defeat defence in depth at subsequent levels. The independence of different levels of defence, i.e. the independence of the features implemented to fulfill the requested functions at different levels, is a key element in meeting this objective.

As the objective of the first level of protection is the prevention of abnormal operation and system failures, if it fails, an initiating event comes into play and a sequence of events is potentially initiated. Then the second level of protection will detect the failures or control the abnormal operation. Should the second level fail, the third level ensures that the safety functions are further performed by activating specific safety systems. Should the third level fail, the fourth level limits accident progression through accident management, so as to prevent or mitigate severe accident conditions with external releases of radioactive materials. The last level (fifth level of protection) is the mitigation of the radiological consequences of significant external releases through the off-site emergency response.

In this logic, the physical barriers normally considered in LWRs (fuel, cladding, primary circuit and containment) are provisions to confine fission products. Their contribution to safety has to be assessed for each specific concept of reactor and considered in the general safety architecture of the plant.

15.1.2 The Fundamental Safety Functions

The objective of the safety approach is to provide adequate means:
- to maintain the plant in a normal operational state;
- to ensure the proper short term response immediately following a postulated initiating event (PIE);
- and to facilitate the management of the plant in and following any design basis accident, and following any plant states beyond the design basis that may occur (i.e. the "severe plant conditions").

To ensure safety (i.e. to meet allowable radiological consequences during all foreseeable plant conditions), the following fundamental safety functions shall be performed in operational states, in and following a design basis accident and in and after the occurrence of severe plant conditions:
- Control of the reactivity;
- Removal of heat from the core; and
- Confinement of radioactive materials and control of operational discharges, as well as limitation of accidental releases.

The possible challenges to the safety functions are dealt with by the provisions (Inherent characteristics, safety margins, systems, procedures) of a given level of defence.

15.3 CURRENT SAFETY APPROACH

Operating nuclear power plants are largely designed following a safety architecture dictated by the implementation of the strategy of defence in depth (physical barriers and levels of defence) as discussed above. In the majority of the plants of the current generation the application of defence in depth is mainly based on deterministic considerations. This means that the plant is deterministically designed against a set of normal and postulated accident situations according to well established design criteria in order to meet the radiological targets. The adequacy of the defence in depth is established by the number of barriers and number and quality of systems in each level of defence.

The current design approach has been shown to be a sound foundation for the safety and protection of public health, in particular because of its broad scope of accident sequence

Safety Approaches in Reactor Design

considerations, and because of its many conservative assumptions which have the effect of introducing highly conservative margins into the design that, in reality, give the plant the capability of dealing with a large variety of sequences, in some cases well beyond those included in the design basis.

The deterministic approach is complemented by probabilistic evaluations with the main purpose of verifying that the design is well balanced and there are no weak areas or systems that could allow for the possibility of high risk sequences. Probabilistic safety assessment is recognized as a very efficient tool for identifying those sequences and plant vulnerabilities that require specific additional preventive or mitigative design features.

Figure 15.1 shows in a very schematic fashion the curve of the target risk that separates acceptable and unacceptable situations (frequency of the event × consequences) and the integration of the level of defences with the probability associated to each event. The success criterion for each level of defence is represented by the area limited by the maximum acceptable consequence and probability for that level. (e.g. dotted area for Level 2). An event sequence is initiated if a challenge (internal or external to the plant) breaks the first level of defence (prevention of abnormal operation and failures). The representation of Fig. 15.1, with adequate values of consequences and probabilities on the axes of the diagram gives a visual representation of the contribution of each level of defence to the general safety of the plant, provides a metric and allows for comparisons of the safety and implementation of defence in depth in different concepts.

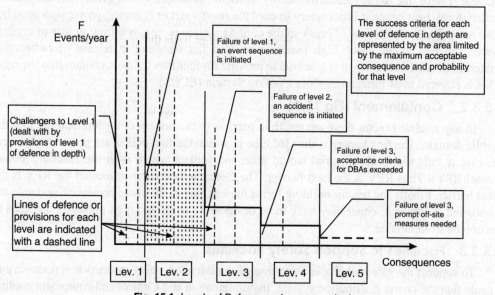

Fig. 15.1: Levels of Defence and success criteria

15.3.1 Achieving Design Safety in Practice

The most important aspects to be built into the design pertain to facilities to control & shutdown, facilities to cool the reactor and facilities to contain radioactivity release. The philosophy is known as shutdown, cool and contain.

15.3.1.1 Facilities to Control & Shutdown Reactor (Fig 15.2)

A reactor shall not go out of control (an excessive power by an abnormal rate nuclear fission reaction) during operation. Moreover in case of a trouble, it is required to immediately stop the nuclear fission process. Therefore, reactor control systems are required to maintain the nuclear fission rate at a constant level and if necessary, facilitate to shutdown the reactor by immediately inserting control rods.

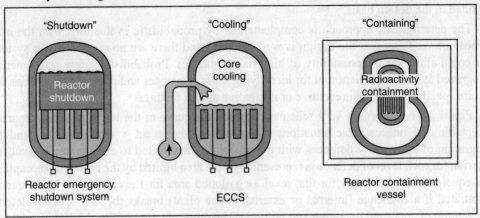

Fig. 15.2: Achieving Design Safety

15.3.1.2 Facilities to cool the reactor

Unlike a coal or oil fired plant, heat continues to be generated even after shutdown of a nuclear reactor due to the radioactive decay of fission products, which generates heat. This is called decay heat hence it is necessary to cool the reactor not only during power operation but also after shutdown. In case of break in the cooling system piping, resulting in loss of cooling water, the fuel would reach high temperatures and fail resulting in release of radioactive material. In such a scenario, it is essential to provide for facilities to inject coolant into the core. This is referred to as Emergency Core Cooling System (ECCS)

15.3.2. Containment (Fig 15.3)

In any nuclear reactor there are multiple barriers to the release of fission material into the public domain. The first barrier is the clad tube in which the fuel pellets are placed and sealed. In case it fails the fissile material would enter the coolant which is in the reactor pressure vessel(RPV). Thus RPV is the next barrier. The Concrete containment around the RPV is the next barrier. Finally the reactor building forms the last barrier. To avoid pressure rise due to the continued decay heat, either they have to be designed to withstand high pressures are need to be cooled from outside.

15.3.3 Facilities to Support Safety Facilities

To support the safe and proper operation of the intended safety systems it is necessary to ensure that the power distribution system, Instrumentation and Control and equipment cooling systems are in place during all states of the plant.

15.3.3.1 Redundancy Diversity and Independence

To ensure that the intended functions of shutdown and cooling are reliably carried out, the nuclear power plant designs use the concepts of redundancy, diversity and independence. Redundancy is a concept in which multiple trains with sufficient capability are provided to the safety equipments and a failure of one train does not cause loss of the safety function. For example, multiple trains of power sources are provided such that failure of even one train would not result in loss of power supply.

Safety Approaches in Reactor Design

Diversity is the design concept to prevent the function loss of multiple facilities simultaneously due to one common cause, by providing different types of facilities. For example, two driving sources of cooling water injection pump systems are to be different such that one is electric driven and the other steam turbine driven, so that injection of cooling water is ensured.

Independency means to maintain independency of the multiple trains so that the safety function is not lost due to a single failure. For example, power sources, control circuits etc. of cooling water injection pumps in multiple systems are designed such that they consist of a power source, detectors and control devices independent from each other and even if one of them were lost, the other cooling water injection pump would be operable.

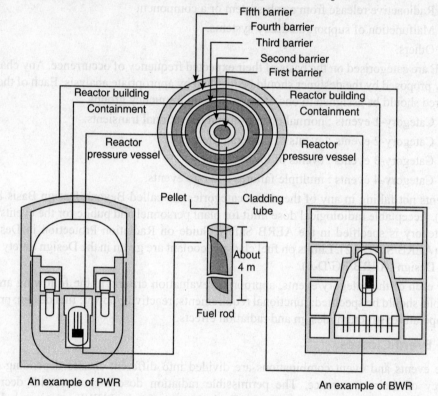

Fig. 15.3: Containment Barriers

15.4 DESIGN BASIS EVENTS (DBE)

Designers must carry out a systematic plant review to identify all postulated design basis events including:
- all process failures resulting from failure of single component or system or combinations
- all serious process failures combined with unavailability of the mitigating systems
- the frequency of such events

Design Basis Events (DBE), which form the basis of design of NPP, include normal operations, operational transients and Postulated Initiating Events (PIE). DBE can be classified on the basis of their consequence and expected frequency of occurrence. Consequences of a rare event can be permitted to be severe while those of a frequent event can be accepted only at very low severity. The first step in analysis is to postulate a number of events affecting process parameters following failure/malfunction of equipment. Each PIE should then be assigned to one of the following groups:-

(i) Reactivity and power distribution anomalies.

(ii) Decrease in primary heat transport (PHT) system inventory.

(iii) Increase in PHT system inventory.

(iv) Increase in heat removal by secondary system.

(v) Decrease in heat removal by secondary system.

(vi) Decrease in PHT system flow rate.

(vii) Radioactive release from a sub-system or a component

(viii) Malfunction of support/auxiliary systems.

(ix) Others.

DBE are categorised on the basis of their expected frequency of occurrence. Any change in category proposed by the designer should be justified by appropriate analysis. Each of the DBE considered should be assigned to one of the following frequency groups.

(i) Category-1 events : normal operation and operational transients.

(ii) Category-2 events : events of moderate frequency.

(iii) Category-3 events : events of low frequency.

(iv) Category-4 events : multiple failures and rare events.

Events not falling in any of the above categories are called Beyond Design Basis Events (BDBE). Acceptable radiological dose limit for plant personnel and public for the events under each category is specified in the AERB Safety Guide on Radiation Protection in Design of PHWR (AERB/SG/D-12). Limits on fuel clad and coolant are given in the Design Safety Guide on Fuel Design (AERB/SG/D-6).

For each of the category/events, appropriate evaluation criteria in the following areas, as applicable, should be specified: functional requirements; reactivity/power; fuel design; pressure and temperature; structural design and radiation effects.

15.4.1 Event Classes

The events and event combinations are divided into different classes depending on the frequency of their occurrence. The permissible radiation dose increases with decreasing frequency keeping the total exposure same. Typical events for the PHWR are given below for illustration

15.4.1.1 Class I-Normal Operation and Operational Transients

Operational process transients such as start-up/shutdown/power changes, expected to occur frequently as part of normal operation and maintenance, are included under this category. Such transients may determine the life of systems /equipment /instrumentation. The frequency of events under this category is expected to be greater than or equal to 1 per reactor-year

- Loss of reactivity control(control rod maloperation)
- Loss of class IV electrical power
- Loss of feedwater flow
- Loss of service water flow
- Loss of Instrument Air

Safety Approaches in Reactor Design

- Loss of Moderator flow
- Fail open of relief valve in primary heat transport system

The above events are expected to occur once or so during plant operation.

15.4.1.2 CLASS II Events of Moderate Frequency

Events of moderate frequency (~ 1 to 10^{-2}) per reactor-year are included in this category.

- Feeder pipe break
- Pressure tube failure
- Flow blockage of fuel channel
- Single heat transport pump seizure
- Service water pipe failure
- Design basis fires

These are expected occur less than once in the lifetime of the plant.

15.4.1.3 Class III Events of Low Frequency

Events of low frequency which are rare events and likely to occur $\sim 10^{-2}$ to 10^{-4} per reactor-year are included in this category.

- Large Loss of Coolant Accident
- Main Steam line Break
- Feedwater pipe break
- Design Basis Earthquake
- Moderator Pipe break

These events are expected to occur less than once per thousand years

15.4.1.4. CLASS IV Multiple Failures and Rare Events

Rare events in this category generally cover multiple failures nsidered important for design and which are likely to occur $\sim 10^{-4}$ to 10^{-6} per reactor-year. For the combination, it is assumed that two independent initiating events, which do not result from a single cause cannot occur simultaneously. Multiple failures considered are based on an initiating event simultaneous with non-availability of a safety system.

- Fuelling machine backing off without replacing reactor closure plug and in turn:
    - Loss of emergency coolant injection
    - Heat transport system isolation failure
    - Failure of crash cooldown of Steam Generators
    - Failure of Dousing
- Main Coolant pump failure

15.4.2 Beyond Design Basis Events

These events are rare events combined with safety system impairment and their expected frequency is less than 1 in 100,000 years.

- Loss of coolant accident (LOCA) plus failure of both the reactor shutdown systems.
- Loss of coolant accident plus failure of emergency core cooling system followed by loss of moderator heat sink.

- Failure of coolant channel seal plug or end fitting leading to ejection of fuel bundle from coolant channel coupled with containment impairment characterized by
  (a) failure of one set of containment Isolation Dampers or
  (b) failure of containment isolation logic or
  (c) one door of main airlock stuck open and seals on second door deflated.

15.5 PROBABILISTIC SAFETY ANALYSIS

Probabilistic Safety Analysis (PSA) is carried out to arrive at a numerical measure of plant risk to the public. It comprises:
- Identification of Potential accidents, calculates the probability of their occurrence and their consequence
- The product of the frequency of postulated accidents and their consequences provided an estimate of plant risk.

PSA develops a mathematical model that relates plant risk to contributory factors such as plant configuration, equipment reliability, operator error probability, operating practices, plant response and system capability.

The elements of PSA are:
- Identification of initiating events
- Event sequence diagrams
- Event tree analysis
- Fault Tree Analysis
- Human reliability Analysis
- Accident sequence Quantification
- Common cause Failure Analysis

Level 1 PSA gives the designer a numerical value of the frequency of individual event sequences that can result in severe core damage. This helps the designer to improve his design. The event tree represents various possible scenarios which can result from the same initiating event. It indicates whether the end point is a stable or damage condition. Operating experience is used to estimate frequency of initiating events. Fault tree analysis identifies the most likely system failure modes, indicating the potential weakness in system design and operation philosophy. Random equipment failure, human errors, test and maintenance unavailability are considered as basic events.

Level 2 PSA is level 1 plus containment performance and source terms (used in calculation of activity release) at containment boundary. This requires severe core damage analysis including core melting, debris formation, progression, thermo hydraulics of core debris, hydrogen production etc..

PSA is a useful assistant to the designer. It ensures adequate redundancy, functional separation and diversity at an early stage. It identifies the risk dominant accident sequences. The designer gets an integrated response of the plant to abnormal events. Identifies important operator actions and test and maintenance programmes. PSA is the most cost effective design tool.

15.6 REGULATORY PROCESS IN INDIA

Over the 60s and 70s the regulatory process for nuclear reactors evolved into five stages of licensing namely Site approval, construction approval, commissioning approval, operating approval and decommissioning approval.

15.6.1 Site Approval

To get the site approval the utility has to submit a report containing detailed documents on
- the site characteristics that could affect the impact of any release of liquid or gaseous radioactive effluents on the natural environment and persons living in the environs of the site
- Site characteristics that might result in external events that in turn might affect the operation of the plant such as seismicity, tornadoes, and flooding, industrial plant, pipeline and transportation failures

And

- Preliminary design report of the full plant.

These reports are reviewed by independent agencies including the central and state governments. Recently, there is need for a public hearing without which site approval is not given.

- Geological, seismological, meteorological characteristics, population distribution, land use, water use, distance from airports and water bodies
- Environmental Clearance-central & state Pollution control boards

15.6.2 Construction Approval

The construction license is given after satisfactory review of:
- Preliminary safety Analysis report (PSAR)-Preliminary Safety Analysis Report-Indicates design basis, Safety Analysis based on postulated initiating events.
- Detailed commissioning specifications and procedures
- Description of the overall quality assurance program governing the design, procurement, manufacture, construction, commissioning and operation of the plant
- Detailed plan for the training of control room and field operators, maintenance and other essential personnel
- Review by Project design safety committee (PDSC) of Atomic Energy Regulatory board(AERB)
- Approval for Construction

Subsequently approval is granted based on the reviews and inspections by AERB committees for every stage such as:
- First pour of Concrete
- Start of erection of Major Equipment
- Commissioning of Coolant & Moderator System with Light Water in case of PHWR
- Hot conditioning of PHT System
- Charging Heavy water into System
- First Criticality
- Reaching 50%,75%,90%, 100% power levels

15.6.3 Operating License

The operating license is a very detailed specification of the operational limits and conditions governing all process systems, safety systems and safety support systems. It defines the envelope for the plant and the licensing basis. In practice, it is often referred to as the technical specifications for operation. Any violation of the technical specification is reported immediately to the regulatory authority.

15.6.4 Regulatory Inspection

- To Ensure that Safety requirements in construction are followed & all pre requisites are fulfilled before going to next stage
- Three to four inspections per year
- Recommendations reviewed by PDSC

The Safety Review Committee for Operating Plants(SARCOP) oversees the plant operation after commissioning. Its responsibilities include:

- Review of reports submitted by utility
- Review of modifications-Hardware, Equipment, Logics
- Implementation of SARCOP Recommendations
- Licensing of Operational Personnel
- Control of Radiation Exposures
- Review of Environmental release reports from Environmental survey labs.

15.7 RADIATION DOSE LIMITS

Strict dose limits are imposed for both occupational workers in the nuclear facilities and the public. A limit of 20mSv per year (averaged over a 5 year period, with no more than 50 mSv in any year) is for occupational workers. For members of the public it is 1mSv per year.

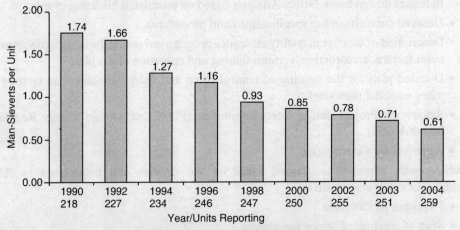

Fig. 15.4: Collective Exposure PWRs

15.8 RADIATION EXPOSURE

The operating experience of different PWRs, BWRs, PHWRS and Gas Cooled reactors is available in open literature. A noteworthy compilation is available of the collective radiation exposure to the operating personnel in the above reactor groups. This does not include the

Safety Approaches in Reactor Design

exposure due to the Three mile island (PWR), Chernobyl (RBMK) and Fukushima (BWR). Figs. 15.4 to 15.7 present the data for a given period. It can be noted that with safety practices and safety culture in the Nuclear Arena with a good regulatory practice, the exposures have reduced considerably with time.

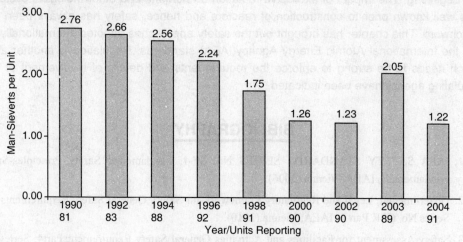

Fig. 15.5: Collective Exposure BWRs

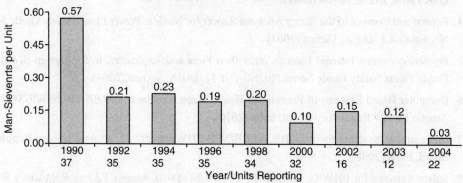

Fig. 15.6: Collective Exposure GCRs

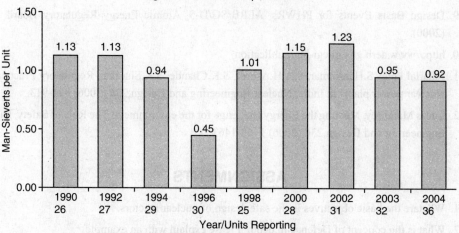

Fig. 15.7: Collective Exposure PHWRs

SUMMARY

Safety in design, construction and operation of nuclear reactors has been in vogue right from the beginning. The impact of excessive radiation on humans and other biological systems was well known prior to construction of reactors and hence, safety has always been the watchword. This chapter has brought out the safety approaches adopted internationally as per the International Atomic Energy Agency(IAEA) standards. Regulation is another area which needs to be strong to enforce the requirements and details of involvement of the regulating agency have been indicated.

BIBLIOGRAPHY

1. IAEA SAFETY STANDARDS SERIES No. SF-1, Fundamental Safety Principles-Safety Fundamentals, IAEA, Vienna (2006).
2. Governmental, Legal and Regulatory Framework for Safety General Safety Requirements Part Series No. GSR Part 1, IAEA, Vienna,(2010).
3. Safety Assessment for Facilities and Activities General Safety RequirementsPart4, Series No. GSR Part 4, IAEA,Vienna,(2009).
4. Format and Content of the Safety Analysis Report for Nuclear Power Plants Safety Guide Series No. GS-G-4.1, IAEA, Vienna,(2004).
5. Protection against Internal Hazards other than Fires and Explosions in the Design of Nuclear Power Plants Safety Guide Series No. NS-G-1.11, IAEA, Vienna,(2004).
6. Computer Based Systems of Pressurised Heavy Water Reactors, AERB/NPP-PHWR/SG/D-25, Atomic Energy Regulatory Board, India (2010).
7. Containment Systems for PHWRS, AERB/NPP-PHWR/SG/D-21, Atomic Energy Regulatory Board, India (2008).
8. Safety Systems for PHWRs, AERB/NPP-PHWR/SG/D-10, Atomic Energy Regulatory Board, India (2005).
9. Design Basis Events for PHWRs, AERB/SG/D-5, Atomic Energy Regulatory Board, India (2000).
10. http://www.aerb.gov.in/cgi-bin/publication
11. Jaharlal koley,S.Harikumar, S.A.H.Ashraf, S.K.Chande, S.K.Sharma,, Regulatory Practices for Nuclear power plants in India, Nuclear Engineering and Design,236,(2006) 894-913.
12. Lucas Mamprey, Meeting the Energy challenge for the environment: The Role of Safety, Nuclear Engineering and Design,236,(2006) 1460-1463.

ASSIGNMENTS

1. Whatare the basic objectives in the safe design of nuclear reactors.
2. What is the concept of Defence in depth. Please explain with an example.
3. What are the different levels of defence in depth.

Safety Approaches in Reactor Design

4. What are the methods of achieving design safety in practice.
5. Clarify the terms Redundancy, Diversity and independence with reference to a nuclear reactor.
6. What are the different stages of approval during the course of the nuclear plant design to operation.
7. What do you mean by Design Basis events. Give examples with reference to a PHWR.
8. What is the difference in the deterministic and probabilistic safety approaches.
9. What are beyond design basis events. How are the consequences of such events managed.
10. Visit the website of the Atomic Energy Regulatory Board, India (www.aerb.gov.in) and collect the list of safety codes and guides issued for the design of the nuclear power plants, fuel cycle facilities and medical usage of radiation sources. Present in a seminar the information so obtained.

Direct Energy Conversion

16.0 INTRODUCTION

Direct conversion of nuclear radiation or the heat energy produced by radiation into electricity has been attempted as it reduces the size of the components. Moreover power sources required for space applications need to be light and have a reasonably long life. Such direct converters would be simple, without any moving parts. Unlike nuclear power plants which produce thousands of megawatts of electrical power, these direct conversion devices typically produce only hundreds of watts to microwatts of electrical power. Nevertheless, these converters have found specialized applications for which low power levels are adequate, such as space satellites, remote meteorological weather stations, and heart pacemakers. This chapter is devoted to a review of such direct conversion concepts.

16.1 THERMOELECTRIC GENERATORS

If two wires or rods, made of different metals, are joined and their junction placed at a different temperature than their opposite ends, a voltage is produced across the unjoined ends. This effect was discovered by Seebeck in 1822; however, because only mill amperes of current at a fraction of a volt are produced by metal wires, the Seebeck effect was used only in thermocouples to measure and control temperatures. Only with the discovery of semiconductors in the late 1950s were materials discovered that could produce useful amounts of electrical power. The basic operation of a thermoelectric cell is shown in Fig. 16.1. Both p-type and n-type semiconductors are used. These semiconductors are made by introducing impurity atoms into the crystal matrix. In a p-type semiconductor, the impurity atoms have fewer valence electrons than the matrix-lattice atoms so that the resulting crystal lattice has positive holes which move easily through the lattice as positive charges. By contrast, in an n-type semiconductor, the impurity atoms have more valence electrons than the lattice atoms. As a result, the lattice has extra free negative electrons. When n-type and p-type materials are joined and the junction heated, the holes and free electrons tend to move away from the hot junction towards the cold junction, much like a heated gas expands and from the hot junction towards the cold junction, much like a heated gas expands and diffuses away from hot regions. This flow of charge, in turn, produces a current through an external load attached to the two cold junctions of a thermoelectric cell. A thermoelectric converter cell is a low-voltage (a few tenths of a volt), high-current (tens of amperes) device. To obtain useful amounts of power and

Direct Energy Conversion

reasonable voltages, several cells are connected together in series to form a thermopile. Because there are no moving parts in thermoelectric cells, they tend to be very reliable (low failure rate). The use of thermoelectric cells has been successfully applied in remote navigational buoys, weather stations, and space satellites and probes.

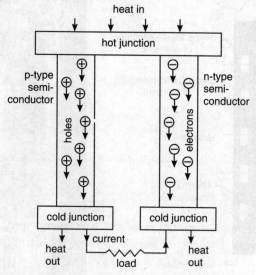

Fig 16..1: Thermoelectric Converter

16.1.1 Radionuclide Thermoelectric Generators

Any high-temperature thermal energy source can be used for a thermopile. One source is the decay heat from radio nuclides. The layout of such a radioisotope thermoelectric generator (RTG) is shown in Fig. 16.2. Many RTGs of varying designs and power capacities have been made and tested in the last 40 years. For example, in Fig. 16.3, the SNAP-7B RTG is shown. The acronym SNAP stands for **S**ystem for **N**uclear **A**uxiliary **P**ower. This 60-W(e) RTG uses 14 tubular fuel capsules containing pellets of 90Sr-titanate whose radioactive decay (half-life 29.1 y) provides the input thermal energy. Around these central fuel tubes are 1120 pairs of lead telluride thermoelectric cells that convert the thermal energy into electrical energy.

Many other RTGs (the odd numbered SNAP series) have been built and used for a variety of terrestrial and space applications (see Table 16.1). These RTGs had electrical power capacities ranging from a few W(e) up to 285 W(e). Although designed for lifetimes generally of 5 years or less, many of these early deep space probes still continue to operate. For example, Pioneer 10, launched in 1972, carried four SNAP-19 RTGs producing an initial total power of about 165 W(e). The pre-launch requirement for the SNAP-19 was to provide power for two years in space; this was greatly exceeded during the mission. The plutonium-238 has a half life of 87.74 years, so that after 29 years, the radiation being generated by the RTG was at 80% of its intensity at launch. However, steady deterioration of the thermocouple junctions led to a more rapid decay in electrical power generation, and by 2005 the total power output was 65 W. As a result, later in the mission only selected instruments could be operated at any one time. The 11989 Galileo (Jupiter) mission contained two of these GPHS-RTGs, the 1990 Ulysses probe in a large-radius solar polar orbit used one, and the 1997 Cassini Saturn mission used three to produce about 890 W(e) of power at launch (Fig 16.4). Besides the RTGs used on a space mission, many low-power radioisotope heat sources are also used to provide heat to warm critical components and instruments.

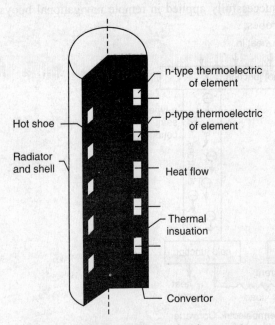

Fig. 16.2: Radioisotope Thermoelectric generator

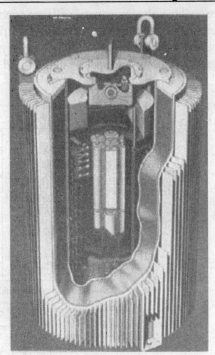

Fig. 16.3: SNAP-7B RTG

Table 16.1: US SNAP Power Generators

SNAP No.	Function	Fuel	Power (We)	Dia. × Ht. (cm)	Mass (kg)	Design Life
3	demonstration	^{210}Po	2.5	12 × 14	1.82	90 d
3A	satellite power	^{235}Pu	2.7	12 × 14	2.10	5 y
-	weather station	^{90}Sr	5	46 × 51	764	2 y min
7A	navigation buoy	^{90}Sr	10	51 × 53	850	2 y min
7B	navigation light	^{90}Sr	60	56 × 88	2100	2 y min
7C	weather station	^{90}Sr	10	51 × 53	850	2 y min
7D	floating weather station	^{90}Sr	60	56 × 88	2100	2 y min
7E	ocean bottom beacon	^{90}Sr	7.5	51 × 53	273	2 y min
7F	offshore oil rig	^{90}Sr	60	56 × 88	2100	2 y min
9A	satellite power	^{238}Pu	25	51 × 24	12	5 y
11	moon probe	^{212}Cm	23	51 × 30	14	90 d
13	demonstration	^{242}Cm	12	6.4 × 10	1.8	90 d
15	military	^{238}Pu	0.001	7.6 × 7.6	0.5	5 y
17	communication satellite	^{90}Sr	25	61 × 36	14	5 y
19	Nimbus weather satellites	^{238}Pu	30	56 × 25	14	5 y
19	Viking/Pioneer missions	^{238}Pu	45			5y
21	deep sea application	^{90}Sr	10	41 × 61	230	5 y

Direct Energy Conversion 221

23	terrestrial uses	^{90}Sr	60	64 × 64	410	5 y
27	Apollo lunar modules	^{238}Pu	60	46 × 46	14	5 y
29	various missions	^{210}Po	500		230	90 d

Data Sources: Mead and Corliss [1966], Furlong and Wahlquist [1999].

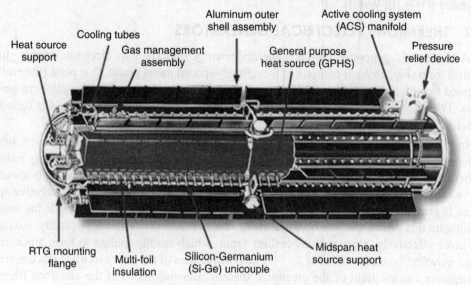

Fig. 16.4: RTG Used on Cassini Probe

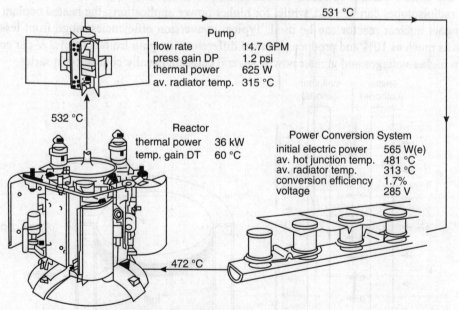

Fig. 16.5: SNAP-10A Power system

16.1.2 Reactor Thermoelectric Generators

The even-numbered SNAP series of thermoelectric devices use a small nuclear reactor to heat a liquid coolant that, in turn, heats the hot junctions of the thermoelectric cells. Such a thermal energy source can provide far more energy than a radionuclide source. Many designs were made but the first nuclear reactor to provide thermal energy for thermoelectric energy conversion was the SNAP-10A system, which was used to provide power to a space satellite launched in April of 1965. The SNAP-10A system is shown in Fig. 16.5. It uses liquid metal NaK as coolant, which is pumped by an electromagnetic pump for which power is supplied by integrated RTGs.[El Wakil]

16.2 THERMIONIC ELECTRICAL GENERATORS

A thermionic generator converts thermal energy directly into electrical energy. In its simplest form (Fig. 16.6), it consists of two closely spaced metal plates. One plate (the emitter) is heated to a high temperature to liberate electrons from its surface into the gap between the plates. The second plate, at a much lower temperature, collects the electrons and is called the collector.

In this manner, a potential difference is developed between the two plates, which, in turn produces a current through an external electrical load. The minimum thermal energy required to liberate an electron varies from material to material. For example, in tungsten it is about 4.5 eV. As electrons are liberated into the gap between the emitter and collector, a negative space charge is created which inhibits the flow of electrons forcing some back towards the emitter. To mitigate this effect, the gap between electrodes is made very narrow (typically. 0.02 cm), and, more effectively, a gas such as cesium vapor, which readily ionizes to form a plasma, is placed between the electrodes (Fig 16.7). The positive ions of the inter electrode gas counteract the negative electric field of the electrons, thereby allowing more of the electrons liberated from the emitter surface to reach the collector. For low power applications, the decay heat from radioisotopes can be used, while, for higher power applications, the heated coolant from a compact nuclear reactor can be used. Typical conversion efficiencies range from less than 1% to as much as 10% and produce potential differences ranging from 0.3 to 1.2 V per cell. To obtain higher voltages and greater power, several cells are usually connected in series.

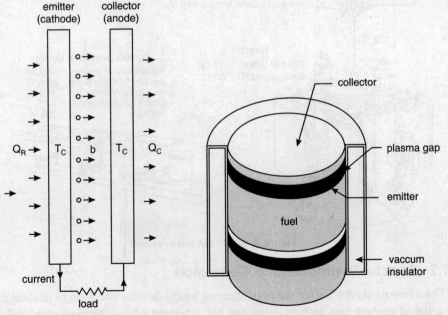

Fig. 16.6: Thermionic Energy Cell   **Fig. 16.7:** Thermionic Radioisotope Generator

16.2.1 Radionuclide Thermionic Generators

Many thermionic generators using the decay heat of radioisotopes have been designed. One of the earliest type of thermionic cells is the so-called isomite battery produced by the McDonnel Douglas Co. (Fig. 16.8). Several prototype isomite cells have been made and tested. These small cells have relatively low emitter temperatures (700 K to 1400 K) and. consequently, low conversion efficiencies of less than 1%. The current densities in these isomite batteries are low (0.1 to 400 niA/cm2) with an output voltage of between 0.1 and 0.15 V to yield power outputs of between 1 arid 20 mW(e). To produce higher temperatures for thermionic cells and thus, greater conversion efficiencies, several efforts have been made to incorporate thermionic cells into a nuclear reactor core. The type of reactor most suitable for in-pile thermionic is a small compact fast reactor using liquid metal as the coolant. Thermal reactors are not suitable because the high temperatures needed for thermionics are not compatible with moderators such as water and beryllium.

Several U.S. and English in-pile experiments have been performed. However, only Russia has produced an in-pile thermionic system that has been deployed in its MIR space programs. Between 1970 and 1984, prototypes were ground tested, and two TOPAZ units were sent into space with the COSMOS satellite. Advanced TOPAZ-II reactors using liquid lithium coolant have been constructed for possible future missions to Mars. The Russian TOPAZ technology has been purchased and tested by the U.S. for evaluation of potential use in its space program. The great advantage of a reactor thermionic generator is the high power levels that can be realized, levels up several kW(e). By contrast, thermionic systems, fueled by decay heat from radioisotopes, usually achieve power outputs of, at most, several tens of watts. Moreover, with UO2 ceramic fuel, very high operation temperature of around 1700 K can be achieved so that conversion efficiencies near 10% can be realized.

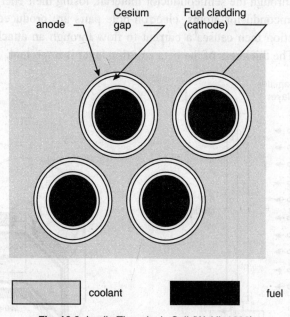

Fig. 16.8: In pile Thermionic Cell (Wakil, 1982)

16.3 AMTEC CONVERSION

A relatively new method for converting thermal energy into electrical energy uses an alkali metal thermal to electric converter (AMTEC). This technology is based on the unique properties of the ceramic β-alumina. This material is a solid electrolyte (BASE) that, while being an electrical insulator, readily conducts sodium ions. When sodium vapor is present at two different pressures separated by this electrolyte, an electrochemical potential is established. Typically, the β-alumina is formed into a cylindrical tube with metal electrodes attached to the inner and outer surfaces. Hot pressurized sodium vapor is fed into the central cavity of the BASE tube where it gives up an electron at the inner electrode (cathode) and the resulting sodium ion Na+ is thermally driven through the electrolyte as a result of cooler lower-pressure sodium vapor on the outside of the tube. As the Na+ ion reaches the outer electrode (anode) it acquires an electron and becomes a neutral sodium atom. The potential difference thus established between the two electrodes can be used to supply electrical power to an external load. The sodium vapor leaving the outer BASE cell surface then diffuses to a cold surface where it is cooled and condensed to a liquid in order to reduce the sodium vapor pressure at the outer surface of the cells. The liquid sodium is then transported back to the heat source where it is vaporized and the cycle is repeated.

16.4 BETAVOLTAIC BATTERIES

In a betavoltaic cell (Fig. 16.9) beta particles emitted by radionuclides deposited on a support plate impinge on a pair of joined n-p semiconductor plates with the n-type plate facing the radioactive source. Such a pair of n-p semiconductors is called a n-p diode. As the beta particles move through the semiconductor material, losing their energy by ionization and excitation of the semiconductor atoms, electron-hole pairs are produced in the diode. Such electron hole production then causes a current to flow through an attached external load as shown in Fig. 16.9. The thickness of the beta emitting layer is important. Too thin and too few

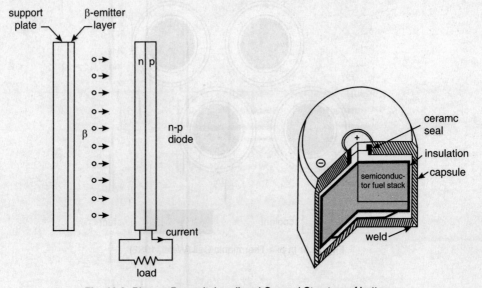

Fig. 16.9: Plannar Betavoltaic cell and General Structure of battery

Direct Energy Conversion

beta particles are emitted. Too thick and many of the emitted beta particles are absorbed or degraded in energy before they can reach the n-p junction. With an optimal thickness of the radionuclide layer, the percent of the emitted beta energy converted into electrical power can be between 1 and 2%. Semiconductors that have been used for beta voltaic cells include Si, Ge, Se, and GaAs. The energy conversion efficiency is about 1%. Although these batteries have small power outputs, they find application in specialized applications, such as biomedical uses to provide power for pacemakers, telemetry, and monitoring devices.

16.5 RADIOISOTOPES FOR THERMAL POWER SOURCES

Although all radioisotopes emit energy when they decay, only handfuls are of practical importance as thermal energy sources for radioisotope power generators. There are several important criteria for the selection of a particular radionuclide as a thermal power source. First, the radionuclide must be available in a stable chemical form that can be encapsulated to prevent any radionuclide leakage into the environment. The radiation emitted by the radioactive decay must be absorbed in the source material, or by a relatively thin layer of surrounding shielding material. to prevent excessive exposures outside the source. The lifetime of the radionuclide must be comparable to or greater than the expected lifetime of the application using the power source. To reduce weight for space applications, the source material must have a high thermal power density. Finally, the cost and availability of the radioisotope must be reasonable.

Table 16.2: Properties of Radionuclide thermal power source

Property	^{144}Ce	^{90}Sr	^{137}Cs	^{147}Pm	^{60}Co	^{242}Cm	^{244}Cm	^{210}Po	^{238}Pu
Half-life	284.9 d	28.84 y	30.07 y	2.623 y	5.271 y	162.8 d	18.101 y	138.4 d	87.7 y
Recov. en. (MeV/dec.)	1.30^a	1.132^a	0.187	0.062	0.09672	6.11	5.803	5.411	5.495
Sp. activity (Ci/g)	3190	136	87.0	927	1131	3307	80.9	4494	17.1
Sp. power (W/g)	24.6	0.916	0.0966	0.341	0.644	120	2.78	144	0.558
1-W activity (Ci/W)	130	149	901	2722	1755	27.6	29.1	31.2	30.7
1-W activity (TBq/W)	4.81	5.52	33.3	101	64.9	1.02	1.08	1.15	1.14
U-Shielding^b	3.6	negl.	3.8	negl.	9.9	neut.	neut.	negl.	neut.
Pb-Shielding^b	6.5	negl.	7.5	negl.	18.0	neut.	neut.	negl.	neut.
Compound	Ce_2O_3	$SrTiO_3$	—	Pm_2O_2	metal	Cm_2O_3	Cm_2O_3	GdPo	PuO_2
Melting pt., °C	1692	1910	—	2350	1495	1950	1950	590	

^{a} In secular equilibrium with its short lived daughter.
^{b} Thickness (cm) to attenuate g-dose rate to 10 rads/h at 100 cm from a 100-W source.

Thus, practical radionuclide thermal power sources should have a half-life of at least several years, low gamma-ray emission, a power density of at least 0.1 W/g, a relatively low cost, and desirable chemical and physical properties (such as stability, chemical inertness, high melting point, etc.). These constraints limit the selection from the hundreds of different radio nuclides to the nine practical radio nuclides listed in Table 16.2. Four of these are beta-particle emitters and can be easily recovered by chemically extracting them from the fission products in the spent fuel produced by nuclear reactors. Four are alpha-particle emitters that must be produced in nuclear reactors by non-fission reactions and, generally, are of low concentration in spent fuel and, consequently, more expensive. The ninth useful radioisotope is ^{60}Co which is plentiful (produced easily by neutron activation of stable ^{59}Co in a reactor). However, ^{60}Co emits energetic gamma photons (1.17 and 1.33 MeV) and thus requires considerable shielding to reduce external exposure.

The two radio nuclides ^{90}Sr and ^{238}Pu have been used the most extensively. ^{90}Sr is inexpensive and plentiful, has a long half-life (28.8 y), and, as a pure beta emitter, requires very little shielding. Because ^{90}Sr has an affinity for bone, it is important that it be well encapsulated to prevent its leakage into the biosphere. The chemical form chosen for ^{90}Sr is usually strontium titanate ($SrTiO_3$) because it is insoluble in water, has a high melting point, and is resistant to shock. Most terrestrial radionuclide sources used to date employ ^{90}Sr.

For space applications, solar cells have been a primary electrical energy conversion device for long, near-earth missions. However, radioisotope thermoelectric generators (RTG) have distinct advantages for certain types of space missions. RTGs do not deteriorate like solar cells when passing through the radiation belts that surround the earth. They are also well suited for deep space missions where solar energy is weak, for moon applications for which solar cells would require large heavy batteries during the long periods of darkness, and for planetary atmospheric probes. Most space applications of RTGs have used ^{238}Pu in the form of plutonium dioxide (PuO_2). This radionuclide has a higher power density than ^{90}Sr and hence a smaller mass of ^{238}Pu is needed for the same thermal power output. Its longer half-life (88 y) also makes it more suitable for long duration space probes. However, ^{238}Pu is much more expensive than ^{90}Sr.

SUMMARY

There is need for electrical power sources with long life especially for space propulsion and the instruments on spacecrafts, as they move away from the earth. Much work has been done in this area and successful power sources developed after longtime research. Solar power supported by direct conversion of nuclear energy provide the answer to this need. This chapter has made a compilation of the various techniques of direct conversion of nuclear energy into electricity.

BIBLIOGRAPHY

1. J.Kenneth Shultis, Richard E. Faw, Fundamentals of Nuclear Science and Engineering, CRC Press; 1 edition (2002).
2. www.world-nuclear.org/info/inf82.html

3. Susan S. Voss, SNAP Reactor Overview, Report AFWL-TN-84-14, Air Force Weapon Laboratory, New Mexico (1984).
4. Steele, O.P., Development of the Control System for SNAP Reactors, IEEE transactions on Nuclear Science, (1966).
5. M.M. El-Wakil, Nuclear Energy Conversion, International Textbook, Scranton, PA, 1982.

ASSIGNMENTS

1. What do you mean by direct conversion of nuclear energy into electricity. What are its advantages.
2. What are the different types of thermoelectric generators. Explain their principle of operation.
3. Write a short note on SNAP power generators.
4. Explain the working of a Thermionic generator. What are its advantages and disadvantages visa-vis Thermoelectric generators.
5. What are the different types of devices converting nuclear radiation to electricity? Explain their principle.
6. Describe the working principle of a Betavoltaic battery.

17

Fusion Energy

17.0 INTRODUCTION

According to Greek mythology, fusion energy, the fire of the Sun, is a gift hard won. When the Titan God Prometheus stole the Sun's fire to give helpless humans the power of survival, Zeus was so angered that he chained Prometheus to a mountain for a thousand years. Today, irresistibly drawn to the challenge of bringing fusion energy down to Earth from the stars, scientists tempt Zeus still. In fact the following reaction is continuously taking place in the sun

$$4\,_1^1H \rightarrow \,_2^4He + 2\,_{+1}^0e.$$

and stars through a complicated chain of events involving Hydrogen and isotopes of carbon, oxygen and nitrogen, referred to as the carbon cycle. It is a slow reaction but produces large amount of heat as we know.

Hans Bethe described the fusion reaction in the stars and was awarded a Nobel Prize. Recently after forty years of strenuous effort, fusion power was finally demonstrated in the laboratory, first in the Joint European Torus in 1991, and then definitively at Princeton University, in a series of experiments that began in December 1993. The next stage involved a design study, involving full participation of the European Community, the USA, Japan and Russia on a large Tokomak, the International Thermonuclear Experimental Reactor, ITER intended to demonstrate all the features, physical and technical, required for a commercial reactor. This chapter deals with the basics of fusion and the present status of fusion power.

17.1 BINDING ENERGY

Assessment of the mass of elements by mass spectrograph has shown that the actual mass is always less than the sum of the masses of protons and neutrons. The difference is referred to as mass defect and is related to the energy binding the particles in the nucleus.

$$\text{Mass defect } m = [Z(m_p + m_e) + (A - Z)m_n\}] - M$$

Hydrogen atom has 1 proton and 1 electron and the weight of 1 proton and 1 electron is the weight of hydrogen atom. Hence the above can be rewritten as

$$\text{Mass defect } m = [Zm_H + (A - Z)m_n] - M$$

m_H is 1.008145 and mn is 1.008986 amu and mass defect can be evaluated for all atoms of different elements.

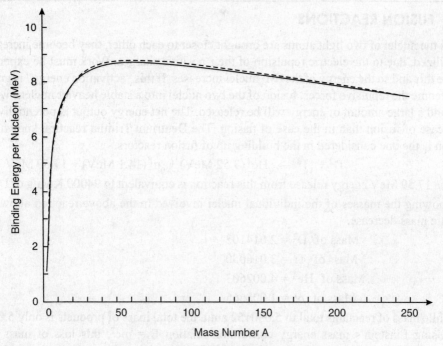

Fig. 17.1: Binding Energy of Different elements

Based on Einstein's special theory of relativity, the mass defect multiplied by c^2, where c is the velocity of light, is a measure of the energy that would be released when Z protons and A-Z neutrons are brought together to form a nucleus. If we are able to give the same amount of energy to a nucleus, it would be able to break the nucleus into protons and neutrons. The energy equivalent of mass defect is called the binding energy of the nucleus.

Taking velocity of light as $2.998*10^{10}$ cm/s the energy E in ergs is obtained as

$$E \text{ (erg)} = m(g)*8.99*10^{20}$$

Converting E to Mev (I Mev is $1.602*10^{12}$ ergs) and m to amu (1amu = 1.66×10^{-24} g) we get

$$E \text{ (Mev)} = 931* m$$

Considering U235, the isotopic mass is 235.1175 and atomic number is 92. The binding energy is calculated as

$$BE/A = 931/235 \, [(1.00814*92) + (1.00898*143) - 235.1175]$$
$$= 7.35 \text{ Mev per nucleon (neutron + proton)}$$

Curve of BE per nucleon for elements of different mass numbers are given in Fig. 17.1. It can be seen that BE per nucleon is low for low mass numbers and it rises to nearly 8.7 Mev for iron (Fe of mass number 56). The BE per nucleon then falls gradually. Thus iron can be said to be the most stable element as it has the maximum BE.

From this curve we can note the following:

Nuclides which are below Fe will have a tendency to become stable by means of fusion process to become a higher mass number element which is more stable. Nuclides which are above would have a tendency to break up or fission to form nuclides with mass numbers close to Fe.

17.2 FUSION REACTIONS

As the nuclei of two light atoms are brought closer to each other, they become increasingly destabilized, due to the electric repulsion of their positive charges. Work must be expended to achieve this and so the energy of the two nuclei increases. If this "activation energy" is provided to overcome the repulsive forces, fusion of the two nuclei into a stable heavier nucleus will take place and a large amount of energy will be released. The net energy output is potentially larger in the case of fusion than in the case of fission. The Deutrium Tritium reaction known as DT reaction is the one considered in the building up of fusion reactors.

$$_1D^2 + {_1}T^3 \rightarrow {_2}He^4 (3.52 \text{ MeV}) + {_0}n^1 (14.1 \text{ MeV}) + 17.59 \text{ MeV}$$

The 17.59 MeV energy release from this reaction is equivalent to 94000 Kwh/g of DT fuel.

Knowing the masses of the individual nuclei involved in the above reaction allows us to calculate mass decrease.

Mass of $_1D^2$ = 2.014102

Mass of $_1T^3$ = 3.016050

Mass of $_2He^4$ = 4.002603

Mass of $_0n^1$ = 1.008665

While mass of reactants total to 5.030152 amu, the total mass of products is only 5.011268 amu. Using Einstein's mass energy conversion relation $E = mc^2$, this loss of mass would appear as 17.6 MeV/nucleus, of energy in the DT fusion reaction. Since the nucleus contains 2 Deutrium atoms, the energy release would be 8.8 MeV/Nucleon of Deutrium. This energy is an order of magnitude higher than in fission of U235.

Deuterium and tritium are the main ingredients in most fusion reactions. Deuterium is a stable form of hydrogen; it is found in ordinary water. Tritium is a radioactive form of hydrogen, not found in nature. In contrast to the situation with fission, where tritium is produced (and thus contributes to radioactivity), here it is consumed.

17.3 FUSION FUEL AVAILABILITY

Fusion fuels are cheap and uniformly distributed on Earth. Seawater contains deuterium D in practically inexhaustible quantities. Deutrium can be extracted from sea water through electrolysis or distillation. The processes are well known with the development of D_2O (Heavy Water) which is used in Pressurised Heavy water reactors. Tritium, a radioactive isotope of hydrogen with a short half life of 12.33 years, hardly occurs in nature. It can, however, be produced in a power plant from lithium, which is likewise abundantly available. Natural Lithium occurs as 7.5% of $_3Li^6$ and 92.5% as $_3Li^7$ isotope. An exothermic reaction can occur by thermal neutrons with the first isotope and an endothermic reaction by Fast neutrons with the second isotope as given below.

$$_3Li^6 + {_0}n^1(\text{thermal}) \rightarrow {_2}He^4(2.05 \text{ MeV}) + {_1}T^3(2.73 \text{ MeV}) + 4.78 \text{ MeV}$$

$$_3Li^7 + {_0}n^1(\text{fast}) \rightarrow {_2}He^4 + {_0}n^1 + {_1}T^3 - 2.47 \text{ MeV}$$

As shown above, tritium can be obtained from lithium, Li-6, a relatively abundant metal found in mineral ores. A simple calculation, based on the fact that there is one deuterium atom in every 6500 atoms of hydrogen, shows that, in 65,000 Kg of water there is about 1 Kg of deuterium. Now, water is in general an abundant resource on our planet. This fact, together with the fact that enormous amounts of energy are released in fusion reactions, makes fusion an essentially non depletable energy source.

Fusion Energy

Some of the fusion reactions are depicted in Table 17.1.

Table 17.1: FUSION REACTIONS AND THEIR ENERGY RELEASE

Reactants	Branching Ratios [percent]	Products	Energy yield [MeV]	Energy yield [Joule]
$_1D^2 \mid _1D^2$ (proton branch)	50	$_1T^3$ (1.01 MeV) $\mid _1p^1$ (3.02 MeV)	4.03	6.460×10^{-13}
$_1D^2 \mid _1D^2$ (neutron branch)	50	$_2He^3$ (0.82 MeV) $\mid _0n^1$ (2.45 MeV)	3.27	5.240×10^{-13}
$_1D^2 \mid _1T^3$		$_2He^4$ (3.54 MeV) $\mid _0n^1$ (14.06 MeV)	17.60	2.818×10^{-12}
$_1D^2 \mid _2He^3$		$_2He^4$ (3.66 MeV) $\mid _1p^1$ (14.6 MeV)	18.3	2.930×10^{-12}
$_1T^3 \mid _1T^3$		$_2He^4 \mid 2\,_0n^1 \mid$ 11.3 MeV	11.3	1.810×10^{-12}
$_2He^3 + _1T^3$	51	$_2He^4 + p + _0n^1 + 12.1$ MeV	12.1	1.940×10^{-12}
	43	$_2He^4$ (4.8 MeV) + $_1D^2$ (9.5 MeV)	14.3	2.292×10^{-12}
	6	$_2He^5$ (2.4 MeV) + $_1p^1$ (11.9 MeV)	14.3	2.292×10^{-12}
$_1p^1 + _3Li^6$		$_2He^4$ (1.7 MeV) + $_2He^3$ (2.3 MeV)	4.02	6.440×10^{-13}
$_1p^1 + _3Li^7$	20	$2\,_2He^4 + 17.3$ MeV	17.3	2.773×10^{-12}
	80	$4Be^7 + _0n^1 - 1.6$ MeV	−1.6	-2.565×10^{-13}
$_1D^2 + _3Li^6$		$2\,_2He^4 + 22.4$ MeV	22.4	3.391×10^{-2}
$_1p^1 + B^n$		$3\,_2He^4 + 8.7$ MeV	8.68	1.390×10^{-12}
$_0n^1$ (thermal) + $_3Li^6$		$_2He^4$ (2.05 MeV) + $_1T^3$ (2.73 MeV)	4.78	7.660×10^{-13}
$_0n^1$ (fast) + $_3Li^7$ (endothermic)		$_2He^4$ (2.1 MeV) + $_1T^3$ (2.7 MeV)	−2.47	-3.96×10^{-13}
$_2He^3 + _2He^3$		$_2He^4 + _1p^1 + _1p^1$	12.9	2.065×10^{-12}

17.4 ISSUES IN FUSION

The basic challenges of fusion are the following:

(a) heating of the reacting mixture to a very high temperature, to overcome the repulsive forces of positively charged nuclei;

(b) compressing the mixture to a high density so that the probability of collision (and thus reaction) among the nuclei can be high; and

(c) keeping the reacting mixture together long enough for the fusion reaction to produce energy at a rate that is greater than the rate of energy input (as heat and compression).

The first challenge is that of providing a huge amount of energy to the reactants. This is why fusion is called a thermonuclear reaction. Table 17.2 shows the mind-boggling temperature thresholds for the Different fusion reactions. The D.T reaction requires the least temperature, out of all reactions and hence has been pursued for research. At the Tokomak reactor in Princeton, NJ, a record-breaking one-second 10.7 MW burst was achieved with a 50-50 deuterium-tritium fuel.

Table 17.2: Heating requirements for selected fusion reactions

	Fusion Reaction		Threshold Temperature (°C)
D + D	= $_2He^3$ + n	+ 3.3 MeV (79 MJ/g)	400,000,000
D + D	= T + p	+ 4.0 MeV (97 MJ/g)	400,000,000
D + T	= $_2He^4$ + n	+ 17.6 MeV (331 MJ/g)	45,000,000
D + $_2He^3$	= $_2He^4$ + p	+ 18.3 MeV (353 MJ/g)	350,000,000

D = deuterium; T = tritium; p = proton; n = neutron.

17.5 THERMONUCLEAR REACTION IN PLASMA

With increasing temperature, all materials are successively transformed from the solid to the liquid and then to the gaseous state. If the temperature is further increased, one gets plasma, the fourth state of matter (Fig 17.2). The atoms of the gas decompose into their constituents, electrons and nuclei. The name plasma was introduced by Langmuir thinking of an analogy to the plasma of a biological cell surrounded by a membrane as the plasma sheath.

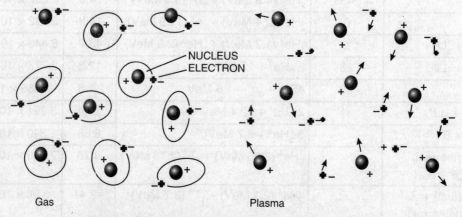

Fig. 17.2: Generation of Plasma through heating the gas nuclei

The Plasma consists of a highly ionized gas in an electrical discharge created by the acceleration of electrons. Equal number of electrons and positively charged ions are present making the medium electrically neutral. Through injection of energy into the plasma its temperature can be raised and particles like deuterons reach the speed for favorable fusion. The term thermonuclear is used for reactions induced by high thermal energy.

The plasma must be created and heated to the necessary temperature and kept together. The presence of solid vessels is ruled out because they would carry away the heat and not allow the plasma to reach the ignition temperature. This challenge is referred to as the confinement problem. Magnets (magnetic confinement) and lasers (inertial confinement) are used instead (in designs that are too complicated to concern us here) to confine the plasma.

17.5.1 Inertial Confinement Fusion (ICF)

One approach to achieving the necessary conditions for fusion on Earth is to exploit the inertia or mass of the fusing nuclei. Inertial fusion involves the firing many times per second of high energy particle or laser beams from all directions at tiny solid fuel pellets in a reaction chamber (Fig 17.3). Material ablated off the pellet by the high energy beams drives a shock wave towards the pellet centre through an inverse rocket or implosion process, raising its temperature and density. This implosion leads to sufficient fusion reactions occurring to overcome the losses, and a large amount of energy is released in a micro-explosion. The resulting alpha particles, neutrons, and radiation flows radially out towards the reaction chamber walls. These are situated far enough typically meters away and built so as to be able to withstand the shock loading. The major experiment worldwide is the National Ignition Facility (NIF) at the Lawrence Livermore National Laboratory (LLNL) in the USA.

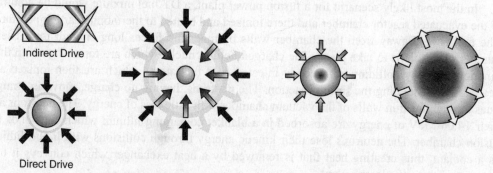

Fig. 17.3: Steps of Inertial Confinement using fusion pellets (LLNL)

17.5.2 Magnetic Confinement

The properties of plasma are very different from those of ordinary gases. Plasma is, for example, electrically conductive. Its motion can therefore be influenced by electric and magnetic fields. This property is exploited in fusion devices by confining the hot plasma in a magnetic bottle, thereby keeping it away from the material walls Fig. 17.4. This feature forms the basis of magnetic confinement fusion, which has been under investigation since the 1950s.

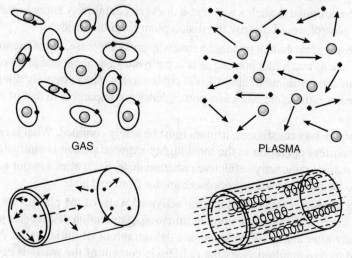

Fig. 17.4: Plasma Confinement using Magnetic field

17.6 LAWSON'S CRITERIA

In addition to providing a high temperature requirement to enable particles to overcome the coulomb barrier, the temperature must be maintained for a sufficient confinement time and with a sufficient ion density in order to obtain the energy of fusion reaction. The overall conditions which must be met are usually stated in terms of the product of the ion density and confinement time referred to as Lawson criterion. It requires that the product of particle density (in nuclei per cubic centimeter) and confinement time (in seconds) for a DT reaction must exceed $10^{14} s/m^3$. For a DD reaction it must be greater than $10^{16} s/m^3$. This criterion can be satisfied, for example for DT reaction, by having 10^{14} nuclei/cm^3 held together for one second.

17.7 FUSION POWER PLANTS

In the most likely scenario for a fusion power plant, a DT fuel mixture would be admitted to the evacuated reactor chamber and there ionized and heated to thermonuclear temperatures. The fuel is held away from the chamber walls by magnetic forces long enough for a useful number of reactions to take place. The charged helium nuclei which are formed give up their kinetic energy by colliding with newly injected cold fuel atoms which are then ionized and heated, thus sustaining the fusion reaction. The neutrons, having no charge, move in straight lines through the thin walls of the vacuum chamber with little loss of energy. The neutrons and their 14.06 MeV of energy are absorbed in a blanket containing lithium which surrounds the fusion chamber. The neutrons lose their kinetic energy through collisions with liquid lithium as a coolant, thus creating heat that is removed by a heat exchanger which conveys it to a conventional steam electric plant.

The neutrons themselves ultimately enter into nuclear reactions with the two isotopes of lithium Li6 and Li7 to generate tritium which is separated and fed back into the reactor as a fuel. The successful operation of a fusion power plant will require the use of materials resistant to energetic neutron bombardment, thermal stress, and magnetic forces.

17.8 SAFETY AND ENVIRONMENTAL CONSIDERATIONS

Safety precautions are necessary because of the radioactive tritium and high-energy neutrons, which activate the walls of the plasma vessel. A fusion power plant has a unique feature: It can be designed in such a way that it does not contain any energy sources that, should they get out of control, could destroy the fusion plant from the inside.

A fusion power plant cannot undergo a runaway energy release. In the combustion chamber there is always just as much fuel burning as is actually needed: about one gram of deuterium and tritium distributed in a volume of about 1,000 cubic meters. The extremely diluted fuel, despite the high temperature, therefore has a low power density, comparable to that of an ordinary light bulb.

The radioactive fuel constituent, tritium, must be safely confined. What is released from the power plant in normal operation to the most highly exposed person is equivalent to about one per cent of the natural radioactive effective radiation dose. Even after a major accident the limit values for introducing evacuation procedures are far from reached.

Remaining as radioactive wastes are the activated walls of the plasma vessel, which have to be placed in intermediate storage upon termination of operation of a reactor plant. Its activity decreases rapidly after about 100 years to one thousandth of its initial value. After decay time of one hundred to five hundred years the radiotoxic content of the waste is comparable to the hazard potential of the ash from a coal fired power plant, which contains radioactive substances. If special materials with low activation potential are used, a major part of the materials could be recycled and reused.

17.9 THE "COLD FUSION" CONFUSION

In March of 1989, two chemists called a press conference at the University of Utah to announce a startling discovery which had eluded physicists for decades. They claimed to have

Fusion Energy

produced "fusion in a jar." This reaction was claimed to have occurred at room temperature within a palladium electrode immersed in a beaker of deuterium-containing water:

$$_1H^2 + {}_1H^2 = {}_2He^4 + 23.85 \text{ MeV}$$

This new nuclear reaction – if indeed possible at such extremely low temperatures – would produce a larger amount of energy than the traditional ones. It is not surprising, therefore, that a frenzy or activity followed this announcement. Many months of frantic research activity were spent by scientists, in dozens of laboratories all over the world, to reproduce these results. The scientists themselves got caught up in the media 'show' and there were almost daily claims and counter-claims about the validity of this new approach to harnessing fusion. The final verdict, at least for the time being, was disappointing: the claims were too good to be true.

Bursts of heat were indeed detected, suggesting that some unusual process is taking place within the palladium electrode, but no characteristic byproducts of the possible reactions (neutrons, gamma rays or enough tritium) were detected. There is no question that, if indeed possible, this reaction would rank as one of the major discoveries in the history of mankind and would solve most of world's energy problems. This idea has probably led the two scientists to announce their results before verifying them thoroughly. Society will thus have to continue to seek more complicated – and more expensive – solutions to its energy problems.

17.10 FUSION RESEARCH IN INDIA

Institute for Plasma Research(IPR) - a premier research organization in India is involved in research in various aspects of plasma science including basic plasma physics, magnetically confined hot plasmas and plasma technologies for industrial application. The institute is currently in the process of building a Steady State Superconducting Tokomak (SST-1) and is one of the partners in the international ITER project. Design and engineering of India's first tokamak ADITYA started 1982 and it was commissioned in 1989 and full-fledged tokomak experiments started. A dynamic experimental programme focusing on transport due to edge turbulence has resulted in major discoveries in this field. This period also saw development of new programmes in plasma processing and basic and computational plasma research.

A steady state superconducting tokomak SST-1 is under design and fabrication at the Institute for Plasma Research. The objectives of SST-1 include studying the physics of the plasma processes in tokomak under steady state conditions and learning technologies related to the steady state operation of the tokomak. These studies are expected to contribute to the tokomak physics database for very long pulse operations.

17.11 ADVANTAGES OF FUSION ENERGY

As a source of energy, fusion would offer many advantages:

1. Abundant fuel supply. The major fuel, deuterium D, may be readily extracted from ordinary water, which is available to all nations. The surface waters of the earth contain more than 1012 metric tonnes of deuterium, an essentially inexhaustible supply. The tritium required would be produced from lithium, which is available from land deposits or from sea water which contains thousands of years' supply. The world-wide availability of these materials would thus eliminate international tensions caused by the imbalance in the fossil fuel supplies.

2. Non critical design. In comparison to fission power plants, no risk of a nuclear criticality accident exists. The amounts of deuterium and tritium in the fusion reaction zone will be so small that a large uncontrolled release of energy would be impossible. In the

event of a malfunction, the plasma would strike the walls of its containment vessel and cool itself out.

3. No air pollution. Since no fossil fuels are used, there will be no release of chemical combustion products because they will not be produced.
4. No high level nuclear waste. Similarly, there will be no fission products nor transuranics formed to present a handling and disposal problem. Radioactivity will be produced by neutrons activating the reactor structure, but careful materials selection is expected to minimize the handling and ultimate disposal of the activated materials.
5. No generation of weapons material. Another significant advantage is that the materials and by-products of fusion are not suitable for use in the production of nuclear weapons.

SUMMARY

Bringing fusion to the level of technological viability for electricity production and to commercial scale will take several decades and billions of dollars of further research and development. According to (probably) optimistic estimates, the construction of a commercial plant might be achieved by 2040, but only if this R&D support is increased substantially. Given the tremendous costs involved, international collaboration is being pursued. Design and construction of the International Thermonuclear Experimental Reactor (ITER), which will go beyond short fusion bursts, is being financed by the U.S., Japan, Russia and the European Union; it is expected to cost some $10 billion and the jury is still out regarding its successful completion.

BIBLIOGRAPHY

1. Kenneth Fowler, The Fusion quest, The John Hopkins University Press, London,(1997).
2. M.Ragheb, Fusion Concepts, MIT(OCW), 2011.
3. Jeffrey P. Freidberg, Plasma Physics and Fusion Energy, Cambridge University Press(2012).
4. http://www.iter.org/
5. http://www.ipr.res.in/

ASSIGNMENTS

1. What is the principle of a fusion reactor. How does it differ from the Fission reaction?
2. What are the important steps in realizing a practical Fusion reactor.
3. Sun is a Fusion reactor. Explain this statement.
4. Why is plasma confinement needed. Explain the different types of confinements in practice.
5. Is Fusion reactor having advantages over Fission reactors. Discuss.
6. Collect literature about fusion research activities in USA and Europe and present in a seminar.

APPENDIX

RADIATION AND HEALTH

INTRODUCTION

Radiation and radioactive materials are part of our environment. The radiation in the environment comes from both cosmic radiation that originates in outer space, and from radioactive materials that occur naturally in the earth and in our own bodies. (Fig. 1) Together, these are known as background radiation. Everyone is exposed to background radiation daily. In addition, radiation and radioactive materials are produced by many human activities. Radiation is produced by x-ray equipment and by particle accelerators used in research and medicine. Radioactive materials are produced in nuclear reactors and particle accelerators.

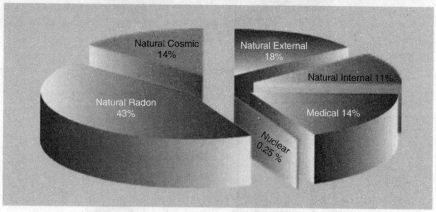

Fig. 1: Radiation from different sources

Today, radiation is a common and valuable tool in medicine, research and industry. It is used in medicine to diagnose illnesses, and in high doses, to treat diseases such as cancer. Also, high doses of radiation are used to kill harmful bacteria in food and to extend the shelf life of fresh produce. Radiation produces heat that is used to generate electricity in nuclear power reactors. Radioactive materials are used in a number of consumer products, such as smoke detectors and exit signs, and for many other research and industrial purposes. (Fig. 2)

This appendix provides answers to general questions in the minds of the reader.

WHAT IS RADIATION?

Radiation is energy that moves through space or matter at a very high speed. This energy can be in the form of particles, such as alpha or beta particles, which are emitted from radioactive materials, or waves such as light, heat, radio waves, microwaves, x-rays and gamma rays. (Fig. 3) Radioactive materials, also known as radio nuclides or radioisotopes, are atoms that are unstable. In nature, there is a tendency for unstable atoms to change into a stable form. As they change form, they release radiation.

Fig. 2: Use of Radiation

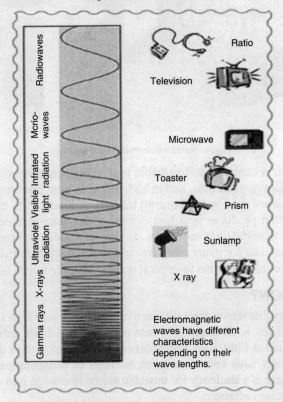

Fig. 3: Sources of Electromagnetic Radiation

Appendix

Radiation that can produce ions when it interacts with matter is called ionizing radiation. Ions are the charged particles that are produced when electrons are removed from their positions in the atoms. Alpha particles, beta particles, x-rays and gamma rays are forms of ionizing radiation. On the other hand, radiation that is not capable of producing ions in matter is known as nonionizing radiation. Radiowaves, microwaves, heat waves, visible light and ultraviolet radiation are forms of nonionizing radiation. This appendix focuses on the health effects of ionizing radiation.

How does ionizing radiation affect health?

Ionizing radiation affects health when it causes changes in the cells of the human body. It does this by breaking the chemical bonds that hold together groups of atoms called molecules. For example, DNA molecules, which contain a person's genetic information, control the chemical and physical functions of human cells. If damaged, the DNA molecules are able to repair. Damage to DNA molecules will affect the ability of the cells to do their work and to pass information to new cells. (Fig. 4)

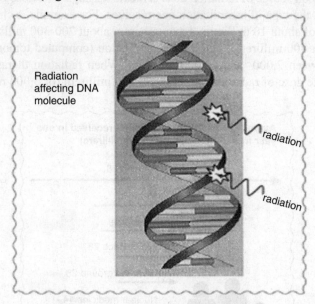

Fig. 4: Ionizing radiation effect on human body

How is the radiation dose measured?

As radiation moves through matter, some of its energy is absorbed into the material. The amount of radiation energy deposited per unit of mass of matter is known as the absorbed dose. The unit of measurement for an absorbed dose of radiation is the rad. When radiation is absorbed by living tissue, the type of radiation, in addition to the absorbed dose, is important in determining the degree of damage that may occur. Alpha radiation, which is heavier and carries more electric charge, causes more damage than beta or gamma radiation. To account for this difference and to give the dose from all types of radiation a common measure, a quantity known as dose equivalent is used. The dose equivalent is found by multiplying the absorbed dose (in rads) by a "quality factor" for the specific type of radiation. The unit for this measurement is called the rem. In many cases, the amount of radiation dose equivalent is much less than one rem. So, a smaller unit, the millirem, is used (1 rem = 1,000 millirem).

What is the dose equivalent in the U.S. from all radiation sources?

In the United States, the annual absorbed dose includes exposure to background radiation, indoor radon and different radiation sources (e.g., industrial and medical). The annual dose to individuals varies by where they live and whether they had medical x-ray or nuclear medicine procedures in the past year. On the average, the dose equivalent in the United States from all sources is about 360 millirem per year. (Fig. 5)

People are exposed to various levels of radon gas which occurs naturally in the air that we breathe. Radon in indoor air results in a dose equivalent of about 200 millirem annually, depending on the radon level in a person's home. Besides radon, the average dose equivalent from background radiation to residents of the United States is about 100 millirem per year. About 40 percent of this dose comes from radioactive materials that occur naturally in our bodies. The rest comes from outer space, cosmic radiation, or from radioactive materials in the ground.

Another common source of radiation dose is medical x-rays. The dose equivalent from this source varies with the type of examination one receives. For example, a chest x-ray results in a dose equivalent of about 10 millirem; a mammogram about 200-300 millirem; an abdominal examination about 400 millirem); and a CT examination (computed tomography, also called "CAT scan"), between 2,000- and 10,000 millirem. When radiation therapy is used to treat cancer, a very large dose of radiation, about 5,000,000 millirem (or 5,000 rem) is delivered to the tumor site.

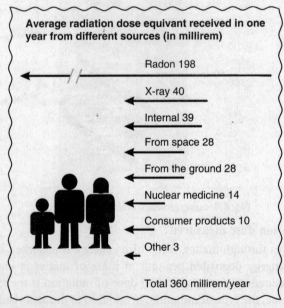

Fig. 5: Average radiation dose in USA

Fallout from nuclear weapons tests is another source of radiation exposure. The exposure from this source has decreased with time since atmospheric testing was stopped in the United States in 1962. Now it contributes less than 1 millirem per year.

What health effects can be caused by radiation?

Exposure to radiation may lead to different health effects. The type and probability of the effects produced generally depend on the radiation dose received.

Appendix

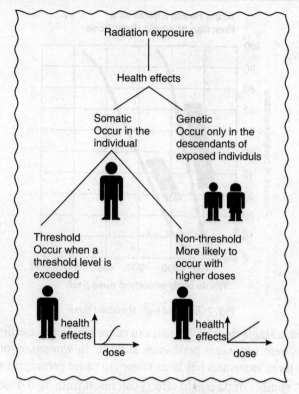

Fig. 6: Health effects of radiation

This could result in health problems for the exposed individuals or in genetic defects that may show up in their descendants. Background radiation is an example of a low dose of radiation and of a low dose rate. A low dose rate occurs when exposure to radiation is spread over a long period of time. Exposure to radiation delivered at low dose rates is generally less dangerous than when the same dose is given all at once because the cell has more time to repair damage to the DNA molecule. Generally speaking, there are two types of health effects from radiation — threshold and non-threshold effects. (Fig. 6)

THRESHOLD EFFECTS

High doses of radiation received in a short period of time result in effects that are noticeable soon after exposure. These are known as threshold effects. A certain dose range must be exceeded before they can occur. These effects include radiation sickness and death, cataracts, sterility, loss of hair, reduced thyroid function and skin radiation burns. The severity of these effects increases with the size of the dose.

Radiation Sickness – At doses of about 60 rem, 5% of exposed individuals may vomit. This increases to about 50% at 200 rem. At doses between 300 and 400 rem and without medical treatment, there is a 50% chance that a person will die within 60 days. With proper medical care, however, some people exposed to 1,000 rem could survive. (Fig. 7)

Cataracts – A single exposure between 200 and 500 rem could cause cataracts (clouding of the lens of the eye). If an exposure took place over a period of months, however, it would take about a total of 1,000 rem to produce a cataract.

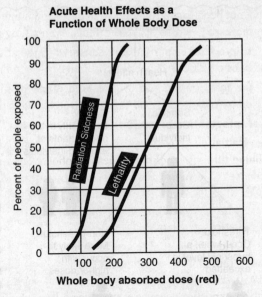

Fig. 7: Effect of high absorbed dose

Sterility – In men, a single dose of 15 rem can cause temporary sterility, and a single dose between 400 and 500 rem can cause permanent sterility. In women, a total dose of 400 rem received over two or three exposures has been known to cause permanent sterility.

Fetal Effects – A number of threshold effects can result from high doses, depending on the stage of development of the fetus. Fetal death is most likely in the first 2 weeks after conception. During this period, a dose of 10 rem may increase the risk of a fetal death by about 0.1 to 1 percent.

Cataracts, malformations, and mental and growth retardation can occur as a result of high radiation doses received 3 to 7 weeks after conception. These effects were not observed at doses of 10 rem or less. Exposures 8 to 15 weeks after conception may lead to mental retardation if the total dose is more than 20 rem. Also, a study of atomic bomb survivors in Japan showed that exposure between 8 and 15 weeks after conception resulted in lower IQ scores in the exposed children. It is estimated that an absorbed dose of 100 rads lowers the IQ score by about 30 points.

NON-THRESHOLD EFFECTS

There are other health effects of radiation that generally do not appear until years after an exposure. It is assumed that there are no threshold doses for these effects and that any radiation exposure can increase a person's chances of having these effects. These are called non-threshold effects. While the chances of these delayed effects occurring increase with the size of the dose, the severity does not. They can occur in the person who receives the radiation dose or in that person's offspring. In the latter case, they are known as genetic effects.

Genetic Effects – No increase of genetic effects from radiation exposure have been found in humans. However, there have been a number of animal studies that have shown that exposure to fairly high doses of radiation increases the chances of genetic effects in the offspring of the exposed animals. They also indicate that radiation does not cause any unique effects. Rather, it increases the number of genetic effects that are normally seen in unexposed animals. Even though genetic effects have not been found to increase in exposed human populations, it is safer to assume that there is an increased chance of these effects, even at low radiation doses.

Fetal Effects – The embryo/fetus is particularly sensitive to radiation. Some studies have shown increases in the rates of childhood cancer in children exposed to radiation before birth.

Cancer – Cancer is the most common non-threshold effect of high radiation doses in humans. The cancers caused by radiation are no different from cancers due to other causes.

What is the source of information on radiation-induced cancer?

It may be noted that Radiation is only one of the causes for Cancer, the others being chemical carcinogens inhaled through smoking, excess alcohol, excess caffeine etc..Estimates of cancer risk for specific doses of radiation are based on many studies of groups of people who were exposed to high doses of radiation. These include the survivors of the atomic bombings in Japan, people who were exposed to radiation for medical reasons, and workers who were exposed to high doses of radiation on the job.

The survivors of the World War II atomic bomb explosions in Hiroshima and Nagasaki were exposed to an average radiation dose equivalent of 24 rem, and their health has been carefully studied since 1950. This study is the most important source of information on the risk of cancer from radiation exposure because it involves a large number of individuals who received whole body radiation exposure. It provides information on the risk of increased cancer to all organs and on the variation of risk with age at the time of exposure.

Are all tissues and organs equally sensitive to radiation?

Radiation has been found to induce cancer in most body tissues and organs. Different tissues and organs, however, show varying degrees of sensitivity. The tissues and organs showing high sensitivity include bone marrow (leukemia), breasts, thyroid glands and lungs. In contrast, there is no clear evidence that radiation causes cancer of the cervix or prostate.

Human tissues and organs ranked by sensitivity to radiation induced cancer

High	Moderate	Low
Bone marrow	Stomach	Brain
Breast (premenapausal)	Overy	Bone
Thyroid (child)	Cobn	Utanus
Lung	Biladder	Kidney
	Skin	Esophagus
		Liver

Are all people at equal risk?

Some people are more sensitive to harmful effects of radiation than others. There are a number of factors that influence an individual's sensitivity to radiation. These factors include age, gender, other exposures and genetic factors.

Age – In general, exposed children are more at risk than adults. Breast cancer risk among women exposed to radiation is greatest among women who were exposed before age 20, and least when exposure occurred after menopause. Also, exposed children are at greater risk of radiation-induced thyroid cancer than adults. (Fig. 8)

Gender – In women, the risk of breast and ovarian cancers from radiation is high, but there is no clear evidence that radiation causes breast or prostate cancers in men. Females are also seen to have more radiation induced thyroid cancer than males.

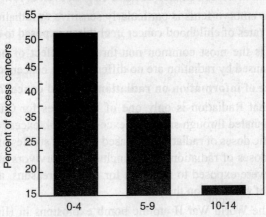

Fig. 8: Effect of age on childhood cancers

Other Exposures – Underground miners exposed to high levels of radon have increased risk of lung cancer, and those who smoke have an even greater risk. Exposure to ultraviolet radiation from the sun following the use of x-rays to treat scalp ringworm conditions increases the risk of developing skin cancer in the area of the skin exposed to both types of radiation.

Genetic Factors – Individuals with certain pre-existing genetic diseases have increased sensitivity to radiation, especially if they receive radiation therapy. For example, children genetically predisposed to cancer of the retina (retinoblastoma) and who are treated with radiation are at increased risk of developing bone cancer following treatment. Patients with ataxia telangiectasia (AT), a rare genetic disorder, are unusually sensitive to tissue damage from radiation therapy, but there is no clear evidence that they are at increased risk of radiation induced cancer.

What is the risk of death from cancers caused by radiation exposure?

In the 1990 report Health Effects of Exposures to Low Levels of Ionizing Radiation – BEIR V, The National Research Council estimated that the excess lifetime risk of fatal cancer following a single exposure to 10 rem would be 0.8 percent. This means that if 1,000 people were exposed to 10 rem each, 8 would be expected to die of cancer induced by the radiation. These deaths are in addition to about 220 cancer deaths that result from other causes. If the 10 rem were received over a period of weeks or months, the extra lifetime risk could be reduced to 0.5 percent. These risk factors are average values for a population similar to that of the United States.

These percentages are not precise predictions of risk, especially at low radiation doses and dose rates. At doses comparable to natural background radiation (0.1 rem per year), the risk could be higher than these estimates, but it could also be lower or even non-existent.

Is there a cancer risk from any radiation exposure?

The risk of increased cancer incidence is well established for doses above 10 rem. For low doses, it has not been possible to scientifically determine if an increased risk exists, but many scientists believe that small doses of radiation do lead to increased cancer risk, and that the degree of risk is directly proportional to the size of the dose. Risk estimates from low doses are obtained by extrapolation from high dose observations.

Appendix

How is the public protected from radiation risks?

Because of the potential for harm from exposure to radiation, radiation protection programs are designed to protect both workers and the general public, their descendants and the environment, while still allowing society to benefit from the many valuable uses of radiation.

Current radiation protection systems are based on the following principles:

- The benefit must outweigh the risk. Radiation exposure must produce a net positive benefit.
- The amount of exposure must be limited. Radiation doses to individuals cannot exceed limits set by state and federal agencies.
- All radiation exposures and releases to the environment must be kept as low as reasonably achievable, below regulatory limits.

What can an individual do to reduce radiation exposure?

Exposure to indoor radon contributes a large portion of the total average dose. Measurements of radon in New York State homes made since 1985 have identified many areas with elevated indoor radon levels. Exposure can be reduced by testing the home for radon and implementing measures to reduce radon levels, if necessary.

Also, a person should receive only x-ray examinations that his or her health care provider thinks are truly necessary for an accurate diagnosis. Alternative, non–x-ray tests should be used instead, if available. However, one should not refuse an x-ray examination that a doctor feels is necessary.

Are there new areas of research that will add to our knowledge of radiation risks?

Recent advances in molecular genetics and microbiology have increased our understanding of cancer development. It is hoped that further research will provide additional information on the risk of radiation-induced cancer and genetic effects, especially at low doses.

GLOSSARY

Absorbed Dose is the amount of radiation absorbed in matter measured in terms of energy per unit mass. The unit traditionally used for absorbed dose is the "rad," but a new unit called a "gray" has been introduced for international use and will eventually replace the rad. One gray equals 100 rads.

Alpha Particles are charged particles that are emitted from some radioactive materials such as radium and radon. The electric charge and mass of the alpha particle are the same as those of the nucleus of a helium atom.

As Low As Reasonably Achievable means making every reasonable effort to keep exposures to radiation as far below as practical the dose limits set in the regulations, taking into consideration the state of technology and other societal and economic considerations.

Atoms are the smallest particles of chemical elements that cannot be divided or broken up by chemical means. Each atom has a large nucleus that contains protons and neutrons, and carries a positive charge equal to the number of protons. Orbiting the nucleus are negatively charged electrons equal in number to the protons in the nucleus.

Background Radiation is radiation that results from natural sources. This includes cosmic radiation and naturally-occurring radioactive materials in the ground and the earth's atmosphere including radon.

Beta Particle is a charged particle emitted from the nucleus of a radioactive material. Beta particles have an electric charge and mass that are equal to those of an electron.

Cosmic Radiation is radiation that originates in outer space and filters through the earth's atmosphere DNA is deoxyribonucleic acid. This is a molecule that controls the chemical and physical functions of cells.

In DNA, a cell passes on information to subsequent generations of cells (daughter cells).

Dose Equivalent is the absorbed dose (in rads or Grays) multiplied by a "quality factor" for the type of radiation in question. This factor is used because some types of radiation are more biologically damaging than others. The unit of dose equivalent is the rem if the absorbed dose is measured in rads. However, if the dose is measured in Grays, a new unit of dose equivalent called the Sievert is used. One Sievert equals 100 rem.

Dose Rate is the radiation dose received over a specified period of time.

Electrons are negatively charged particles that determine the chemical properties of an atom.

Fallout is the airborne radioactive debris from nuclear weapons explosions that has been deposited on the earth.

Gamma Rays are electromagnetic energy that is emitted by a radioactive material. Gamma rays have no mass or electric charge.

Genetic Effects are changes in cells in either the sperm or the egg that are passed on to children born to the exposed individual. If the child survives and becomes a parent, the genetic fault will be passed on to future generations.

Ionizing Radiation is radiation that removes electrons from the atoms it meets, causing them to become electrically charged ions.

Ions are charged atoms that have too few or too many electrons compared to the number of protons, resulting in a positive or negative charge.

Millirem is a unit of measurement equal to 1/1,000th of a rem (1 rem = 1,000 millirem).

Molecules are groups of atoms held together by chemical forces.

Nonionizing Radiation is radiation that is not capable of removing electrons from the atoms it encounters. Examples include visible, ultraviolet and infrared light, and radio waves.

Rad (radiation absorbed dose) is the unit of measurement for the amount or dose of ionizing radiation absorbed by any material.

Radiation is energy that moves through space or matter at a very high speed.

Radioactive Material is material that contains an unstable atomic nucleus and releases radiation in the process of changing to a stable form.

Radium is a naturally occurring radioactive element that occurs in soil. Discovered by Marie and Pierre Curie in 1898, radium was used in radiation therapy and in luminous paints.

Radon is a naturally occurring radioactive element that is produced when radium decays. Radon exists as a gas and is present in the soil. It seeps into the air and can become concentrated in enclosed spaces such as houses. A high radiation dose can result when air containing radon is inhaled.

Rem (roentgen equivalent man) is the unit of measurement for the dose equivalent (the dose of ionizing radiation multiplied by a "quality factor").

BIBLIOGRAPHY

1. American Cancer Society, Cancer Facts & Figures, Atlanta, Georgia, 1998.
2. Cobb, Charles E., Jr., Living with Radiation, National Geographic, Vol. 175, No. 4,

April 1989, National Geographic Society, Washington, DC.
3. Hendee, W.R. and Edwards, F.M., Editors, Health Effects of Exposure to Low-Level Ionizing Radiation, Bristol, UK, 1996.
4. Mettler, Fred A., Jr., M.D., and Upton, Arthur C., M.D., Medical Effects of Ionizing Radiation, Second Edition, Philadelphia, PA, 1995.
5. National Academy of Sciences, Committee on the Biological Effects of Ionizing Radiations, Health Effects of Exposure to Low Levels of Ionizing Radiation, BEIR V, National Academy Press, Washington, DC, 1990.
6. National Academy of Sciences, Committee on Health Effects of Exposure to Radon, BEIR VI, National Academy Press, Washington, DC, 1998.
7. National Academy of Sciences, Institute of Medicine, Exposure of the American People to Iodine - 131 from Nevada Nuclear Bomb Tests, National Academy Press, Washington, DC, 1999.
8. National Cancer Institute, Estimated Exposures and Thyroid Doses Received by the American People from Iodine-131 in Fallout Following Nevada Atmospheric Nuclear Bomb Tests, National Institutes of Health, Washington, DC, 1997.
9. National Council on Radiation Protection and Measurements, Report No. 126, Uncertainties In Fatal Cancer Risk Estimates Used In Radiation Protection, NCRP, Bethesda, Maryland, 1997.
10. National Council on Radiation Protection and Measurements, Proceedings No. 18, at Arlington, VA, April, 1996, Implications of New Data on Radiation Cancer Risk, NCRP, Bethesda, Maryland.
11. National Council on Radiation Protection and Measurements, Proceedings No. 21, at Arlington, VA, April, 1999, Radiation Protection in Medicine: Contemporary Issues, NCRP, Bethesda, Maryland.
12. National Council on Radiation Protection and Measurements, Report No. 93, Ionizing Radiation Exposure of the Population of the United States, NCRP Bethesda, Maryland, 1987.
13. National Council on Radiation Protection and Measurements, Report No. 115, Risk Estimates for Radiation Protection, NCRP, Bethesda, Maryland, 1993.
14. Rall, J.E., Beebe, G.W., Hoel, D.G., Jablon, S., Land, C.E., Nygoard, O.F., Upton, A.C., Yallow, R.S., Zeve, V.H., Report of the National Institutes of Health Ad Hoc Working Group to Develop Radioepidemiological Tables, National Institutes of Health: NIH Publication No. 85-2748, Bethesda, MD, 1985.
15. Shleien, B., Slaback, L.A. Jr., Birky, B.K., Handbook of Health Physics and Radiological Health, Third Edition, Edited by Williams and Wilkins, Baltimore, Maryland, 1998.
16. United Nations Scientific Committee on the Effects of Atomic Radiation, Genetic & Somatic Effects of Ionizing Radiation, United Nations, NY, 1986.
17. United Nations Scientific Committee on the Effects of Atomic Radiation, Sources and Effects of Ionizing Radiation, United Nations, NY, 1993.
18. United Nations Scientific Committee on the Effects of Atomic Radiation, Sources and Effects of Ionizing Radiation, United Nations, NY, 1994.
19. United Nations Scientific Committee on the Effects of Atomic Radiation, Sources and Effects of Ionizing Radiation, United Nations, NY, 2000.

Information on radiation and health can be obtained from the following web sites:

Conference of Radiation Control Program Directors : www.crcpd.org
National Cancer Institute : www.nci.nih.gov
National Academy of Sciences : www.nas.edu
National Council on Radiation Protection : www.ncrp.com
Radiation and Health Physics Homepage : www.umich.edu/~radinfo
US Department of Energy : www.doe.gov
US Environmental Protection Agency : www.epa.gov
US Food & Drug Administration : www.fda.gov
US Nuclear Regulatory Commission : www.nrc.gov
Health Physics Society : www.hps.org
International Atomic Energy Agency : www.iaea.or.at
International Radiation Protection Association : http://www.irpa.net/